AUTOTROPHIC NITROGEN REMOVAL FROM LOW CONCENTRATED EFFLUENTS

Study of System Configurations and Operational Features

for Post-treatment of Anaerobic effluents

AUTOTROPHIC NITROGEN REMOVAL FROM LOW CONCENTRATED EFFLUENTS

Study of System Configurations and Operational Features for Post-treatment of Anaerobic effluents

DISSERTATION

Submitted in fulfillment of the requirements of

the Board for Doctorates of Delft University of Technology

and of the Academic Board of the UNESCO-IHE

Institute for Water Education

for the Degree of DOCTOR

to be defended in public on

Monday November 21, 2016 at 12:30 hrs

in Delft, the Netherlands

by

Javier Adrián SÁNCHEZ GUILLÉN

Master of Engineering in Sanitary Engineering, International Institute for Infrastructural, Hydraulic and Environmental Engineering, IHE Delft, the Netherlands

Chemical Engineer, Universidad Autónoma de Puebla, Puebla, México

born in La Chorrera, Panama

This dissertation has been approved by the
promotors: Prof. dr. ir. J.B. van Lier and Prof. dr. D. Brdjanovic
copromotor: Dr. C.M. Lopez Vazquez

Composition of the Doctoral Committee:

Rector Magnificus TU Delft	Chairman
Rector UNESCO-IHE	Vice-Chairman
Prof. dr. ir. J.B. van Lier	Delft University of Technology / UNESCO-IHE
Prof. dr. D. Brdjanovic	UNESCO-IHE / Delft University of Technology
Dr. C.M. Lopez Vazquez	UNESCO-IHE

Independent members:

Prof. dr. ir. S.E. Vlaeminck	Ghent University
Prof. dr. ir. H.H.G. Savenije	Delft University of Technology
Prof. dr. ir. M.C.M. van Loosdrecht	Delft University of Technology
Dr. S. Lücker	Radboud University Nijmegen
Prof. dr. M.D. Kennedy	UNESCO-IHE / Delft University of Technology, reserve member

This research was conducted under the auspices of the Graduate School for Socio-Economic and Natural Sciences of the Environment (SENSE)

Cover illustration by María Bernadette Sánchez Herrera

CRC Press/Balkema is an imprint of the Taylor & Francis Group, an informa business

Published by:
CRC Press/Balkema
PO Box 11320, 2301 EH Leiden, the Netherlands
Pub.NL@taylorandfrancis.com
www.crcpress.com – www.taylorandfrancis.com
ISBN 978-1-138- 03591-1

"Go to the spring,
drink of it and
wash yourself
there"

"Ga drinken
en u wassen
aan de bron"

"Ve a beber
y a lavarte
en la fuente"

"Allez à la
source, boire
et vous y laver"

IC of Mary
(25-02-1858)

Dedicated to the Immaculate Conception,
to my wife Zahira and our daughter María Bernadette; to our Parents, Family and Friends.

Table of Contents

Summary

On a global scale, sewage represents the main point-source of water pollution and is also the predominant source of nitrogen contamination in urban regions. Even though the existing sewage treatment plants for nitrogen removal, the estimation of the future trends of nitrogen pollution caused by sewage have established that the dissolved inorganic nitrogen (DIN) load discharged by rivers into the oceans will increase in the next 20 years affecting severely coastal areas. Sewage has a low nitrogen concentration compared to sludge reject water. Nevertheless, sludge reject water treatment by Anammox process has opened the possibility to apply autotrophic nitrogen removal in the sewage mainstream. In this regard, sewage treatment through a combined system composed of an upflow anaerobic sludge blanket (UASB) reactor and an Anammox reactor offers an attractive option to control inorganic nitrogen contamination.

The present research is focused on the study of the main challenges that need to be addressed in order to achieve a successful inorganic nitrogen post-treatment of anaerobic effluents in the mainstream. The challenges were classified in terms of operational features and system configuration, namely: (i) the short-term effects of organic carbon source, the COD/N ratio and the temperature on the autotrophic nitrogen removal; (ii) the long-term performance of the Anammox process under a low nitrogen sludge loading rate and moderate to low temperatures; (iii) the Anammox cultivation in a closed sponge-bed trickling filter (CSTF) and (iv) the autotrophic nitrogen removal over nitrite in a sponge-bed trickling filter (STF). The UASB effluent characteristics expected in different climates were taken into account for the study of the simultaneous effect of organic carbon source, the COD/N ratio and the temperature on autotrophic nitrogen removal. Batch tests were carried out under three operating conditions, i.e. 14, 22 and 30°C and COD/N ratios of 2 and 6. This was a first approach to assess the feasibility of the Anammox process as a post-treatment step for anaerobic sewage treatment by UASB reactors. The influence of the fractions of organic matter COD were also evaluated. Thus, for each operating condition three substrate combinations were tested to simulate the presence of acetate as readily biodegradable (RBCOD) and starch as slowly biodegradable (SBCOD) organic matter COD. Although the batch tests do not predict long-term reactor performance, the short-term results confirmed that the Anammox activity was strongly influenced by temperature, in spite of the COD source and COD/N ratios applied.

These results suggest that the Anammox process could be applied as a nitrogen removal post-treatment for UASB reactor in warm and moderate climates.

The influence of the nitrogen sludge loading rate (NSLR) on the Anammox process, expressed as g-N/g-VSS·d, was evaluated for more than 1000 days in a sequencing batch reactor (SBR) with granular Anammox biomass. The NSLR applied was lower than the NSLR capacity of the biomass, i.e. 0.284 g-N/g-VSS·d *vs.* 0.059 g-N/g-VSS·d, respectively. The temperature was lowered progressively from 30^{0}C to 13^{0}C. The total nitrogen removal efficiency, specific activity, granular size stability, i.e. average diameter of the granules, biomass concentration and microbial population variations were investigated. When the NSLR applied was extremely low compared to NSLR capacity of the biomass, irrespective of the temperature utilized, the granular size and biomass concentration decreased. Remarkably, the Anammox population was dominated by the same species during the diverse conditions applied in this study, i.e. *Candidatus Brocadia fulgida*, during the whole research period. Furthermore, the total nitrogen removal efficiency increased when the NSLR applied was close to the NSLR capacity of the system (a difference less than 50%) and small granules were present, i.e. average diameter of 152-171 μm. These results provide useful information for the optimization of the NSLR, especially during the start-up period of granular Anammox bioreactors that use granular Anammox inoculum previously acclimatized to a high NSLR.

Sponge-based Anammox reactors were used to assess the feasibility to immobilize and cultivate Anammox bacteria, specifically in closed sponge bed trickling filters (CSTF). Anammox granular biomass was used as inoculum. The CSTF systems operated at 20 and 30^{0}C immobilized satisfactorily Anammox bacteria and allowed their growth. The temperature of 30^{0}C led to a faster reactor stability and higher nitrogen removal capacity in a shorter period of time compared to the CSTF operated at 20^{0}C. The volumetric nitrogen conversion rate was about 1.52-1.60 kg-N/m^3_{sponge}·d with a short hydraulic retention time (HRT) of about 1.05-1.20 h and an average total nitrogen removal efficiency of 74 ±5 % - 78 ±4 %; these results are comparable to some Anammox full-scale reactors. The CSTFs represent an alternative technology and efficient option for DIN post-treatment from sewage by Anammox since it: (i) provide a suitable surface area for biomass growth, (ii) has a high biomass retention capacity and substrate permeability, (iii) is a simple technology with low operational and maintenance requirements and (iv) high total nitrogen removal efficiency.

The possibility of achieving partial nitritation in sponge bed trickling filters (STF) at 30^{0}C under natural air convection was studied in two reactors with different sponge thickness each, i.e. 0.75 and 1.50 cm. This investigation was carried out to explore new possibilities for the application of the sponge-bed Anammox reactors. Activated sludge was used as inoculum. The coexistence of ammonium oxidizing organisms (AOO) and Anammox bacteria was obtained and attributed to the favorable environment created by the reactors' design and operational regimes e.g. dissolved oxygen of 1.5-2 mg-O_2/L and influent pH around 8.0. Thus, autotrophic nitrogen removal over nitrite in the STFs was obtained and proved that a single stage autotrophic nitrogen removal over nitrite using sponge-bed trickling filters under natural air convection is technically feasible. The total nitrogen removal of 52-54% was obtained and both STF reactors showed robustness to the nitrogen loading rate (NLR) fluctuations, a short HRT (1.71-2.96 h) and had a similar nitrogen removal capacity. This bioreactor is a promising technology and could be coupled with a UASB reactor to develop a cost-effective post-treatment system for ammonium removal provided extensive organic matter removal is achieved upstream.

Resumen

A escala global, las aguas residuales municipales son la principal fuente puntual de contaminación hídrica y la causa predominante de polución debida a nitrógeno en las regiones urbanas. Los ríos poseen un papel dinámico en el transporte del nitrógeno contenido en las aguas residuales municipales, el cual es descargado en las costas marinas de las áreas urbanas. El desbalance causado por los compuestos nitrogenados en los ecosistemas de las costas marinas se ha incrementado dramáticamente. La eutrofización de las costas oceánicas de todo el mundo es la consecuencia directa de la descarga del nitrógeno proveniente de las aguas residuales municipales y la hipoxia (zonas marinas sin vida) es su más severo síntoma. Más de 400 casos de costas marinas afectadas por hipoxia han sido documentados.

En la actualidad existen plantas para el tratamiento de las aguas residuales municipales que incluyen la remoción de nitrógeno. Sin embargo, de acuerdo a la estimación a futuro del comportamiento que seguirá la contaminación debida al nitrógeno proveniente de las aguas residuales municipales, existen indicios de que la carga de nitrógeno inorgánico disuelto (NID) vertida por los ríos en los océanos podría incrementarse en los próximos 20 años afectando severamente las costas urbanas.

Regulaciones ambientales más estrictas para las descargas de nitrógeno provenientes de las aguas residuales municipales, han sido implementadas en muchos países como una medida para mitigar la contaminación por nitrógeno. No obstante, la preservación de las zonas costeras urbanas y la presente situación de crisis financiera a nivel mundial demandan el desarrollo de sistemas de tratamiento de aguas residuales ambientalmente amigables, eficientes y de bajo costo. El descubrimiento de una nueva ruta biológica para la remoción de nitrógeno, es decir, el uso de las bacterias Anammox, ha abierto las posibilidades para mejorar y establecer nuevos esquemas de tratamiento en la línea principal de las plantas de tratamiento para aguas residuales municipales.

Este enfoque podría contribuir a aliviar el impacto de la contaminación por nitrógeno en los cuerpos hídricos. Es así que el tratamiento de las aguas residuales municipales mediante el acoplamiento de los procesos de metanogénesis (tratamiento anaeróbico) y remoción autotrófica de nitrógeno (Anammox) se plantea como una alternativa atractiva. Esto se lograría a través de un sistema combinado que estaría constituido por un reactor anaeróbico de flujo ascendente con manto de lodos (RAFAL) y un reactor Anammox.

Considerando las razones expuestas con anterioridad, la presente investigación fue enfocada al estudio de algunos de los desafíos que son necesarios resolver con el propósito de lograr una integración y operación exitosa del sistema RAFAL-Anammox para la remoción del NID en las aguas residuales municipales. Los desafíos abordados fueron clasificados en términos de los aspectos operacionales y la configuración del sistema. Los desafíos bajo investigación fueron: (i) los efectos de la clase de carbono orgánico, la razón DQO/N y la temperatura sobre la remoción autotrófica de nitrógeno; (ii) el estudio a largo plazo del proceso Anammox sometido a una baja tasa de carga de nitrógeno por unidad de biomasa y operado en un rango de temperaturas de moderada a baja; (iii) el cultivo de bacterias anammox en un filtro percolador cerrado de lecho de esponja (FIPCE) y (iv) la remoción autotrófica de nitrógeno sobre nitrito en un filtro percolador con lecho de esponja (FIPE).

Las características esperadas en el efluente del reactor RAFAL bajo diferentes condiciones climatológicas, fueron tomadas en cuenta para el estudio de los efectos simultáneos de la clase de carbono orgánico, la razón DQO/N y la temperatura sobre la remoción autotrófica de nitrógeno. Pruebas tipo lote se llevaron a cabo usando tres condiciones operativas distintas, es decir, temperaturas de 14, 22 y 30^{0}C y razones DQO/N de 2 y 6. Esta fue una primera aproximación para estimar la factibilidad del proceso Anammox como una etapa de post tratamiento durante el tratamiento anaeróbico de las aguas residuales municipales mediante los reactores RAFAL. La influencia de la DQO debida a las fracciones de materia orgánica también fue evaluada. Así, para cada condición operativa, tres combinaciones de sustratos fueron evaluadas para simular la presencia de acetato como materia orgánica fácilmente biodegradable (FBDQO) y almidón como materia orgánica lentamente biodegradable (LBDQO). A pesar de que las pruebas lote no predicen el desenvolvimiento a largo plazo de los reactores, los resultados a corto plazo confirmaron que la actividad anammox fue fuertemente influenciada por la temperatura, sin importar la fuente de DQO o las razones DQO/N aplicadas. Estos resultados sugieren que el proceso Anammox podría ser aplicado como post tratamiento para la remoción de nitrógeno en el efluente del RAFAL en climas moderados y cálidos.

La influencia sobre el proceso Anammox de la tasa de carga de nitrógeno por unidad de biomasa (TCNB) fue evaluada durante más de 1000 días en un reactor secuencial tipo lote (RSL) con biomasa anammox granular. La TCNB aplicada fue menor que la capacidad de TCNB de la biomasa y la temperatura se disminuyó progresivamente de 30^{0}C a 13^{0}C. La eficiencia de remoción de nitrógeno total, la actividad específica, la estabilidad del diámetro promedio de los gránulos, la concentración de la biomasa y las variaciones en la población microbiana fueron investigadas. Cuando la TCNB aplicada fue extremadamente menor comparada con la capacidad de TCNB de la biomasa, el tamaño del gránulo y la concentración de biomasa disminuyeron, independientemente de la temperatura utilizada. Sorprendentemente, la población anammox fue dominada por una sola especie durante las diversas condiciones de este estudio, es decir, *Candidatus Brocadia fulgida* fue la bacteria Anammox dominante durante el período completo de la investigación. Además, la eficiencia de remoción de nitrógeno total se incrementó cuando el valor de la TCNB aplicada es cercano o igual al valor de la capacidad TCNB de la biomasa y cuando el tamaño de los gránulos fue pequeño. Estos resultados proporcionan información útil para la optimización de la TCNB aplicada, especialmente durante el período de arranque de los reactores Anammox que usan inoculo Anammox granular previamente aclimatado a una TCNB alta.

Varios reactores Anammox con lecho de esponja fueron utilizados para estimar la posibilidad de inmovilizar y cultivar bacterias Anammox. Para ello se diseñaron filtros percoladores cerrados de lecho de esponja (FIPCE). Gránulos Anammox fueron usada como inoculo. Los sistemas FIPCE fueron operados a 20^{0}C and 30^{0}C y lograron la inmovilización de las bacterias Anammox permitiendo su crecimiento. La operación del reactor a 30^{0}C condujo a una rápida estabilización del reactor y a una alta remoción de nitrógeno en un período de tiempo más corto en comparación al FIPCE operado a 20^{0}C. La tasa volumétrica de conversión fue cerca de 1.52-1.60 kg-$N/m^{3}_{esponja}\cdot d$ con un tiempo de retención hidráulico (TRH) corto de 1.05-1.20 h. La eficiencia de remoción de nitrógeno total fue de 74 ±5 % a 78 ±4 %, siendo estos resultados comparables a algunos reactores Anammox que operan a escala real. Los FIPCE representan una tecnología eficiente para el post tratamiento del NID en las aguas residuales municipales, ya que: (i) proporcionan un área superficial apropiada para el crecimiento de la biomasa; (ii) tienen una gran capacidad de retención de biomasa y de permeabilidad al sustrato; (iii) es una tecnología simple con bajos requerimientos operacionales y de mantenimiento y (iv) una alta eficiencia de remoción de nitrógeno total.

La posibilidad de lograr la Nitritación parcial en un filtro percolador con lecho de esponja (FIPE) a 30^{0}C bajo convección natural de aire fue estudiada en dos reactores usando en cada uno un espesor diferente de esponja, es decir, 0.75 y 1.50 cm. Esta investigación fue llevada a cabo con el fin de explorar nuevas posibilidades en la aplicación de reactores Anammox con esponja. Lodo activado fue utilizado como inoculo. La coexistencia de Organismos Oxidantes de Amonio (OOA) y bacterias Anammox fue lograda. Esto es atribuido al ambiente favorable creado por el diseño del reactor y a los regímenes operativos aplicados. Por ejemplo, un ambiente con niveles de oxígeno disuelto de 1.5-2 mg-O_2/L y un pH igual a 8.0 en el afluente. La remoción autotrófica de nitrógeno sobre nitrito usando filtros percoladores con lecho de esponja (FIPE), bajo convección natural de aire, es técnicamente factible. Una remoción de nitrógeno total del 52-54% se obtuvo y ambos reactores mostraron: buen desempeño durante las fluctuaciones de la tasa de carga de nitrógeno (TCN), un TRH corto (1.71-2.96 h) y tuvieron una similar capacidad de remoción de nitrógeno. Este reactor biológico es una tecnología prometedora y podría ser acoplado a los reactores RAFAL para el desarrollo de un post tratamiento efectivo a bajo costo destinado a la remoción de amonio, dado que una extensiva remoción de materia orgánica sea lograda en las etapas previas.

Pollution control of dissolved inorganic nitrogen in urban zones through sewage treatment

Contents

1.1. Source, transport, fate and impact of dissolved inorganic nitrogen pollution in urban areas

1.1.1. Source of inorganic nitrogen

Human beings inherently use and influence the global water cycle. For instance, they have been using water for the transportation and disposal of wastes during centuries (Henze *et al.*, 2008; Jørgensen, 2010). Thus, the wastewaters produced by human activities usually contain a wide variety of substances, e.g. organic compounds, pathogens, nutrients, i.e. phosphorus and nitrogen, etc. The occurrence and content of these substances in wastewaters are the result of people's cultural behavior, income, access to drinking water and environmental factors. The deterioration of surface and groundwater quality is among the principal impacts of the anthropogenic intervention on the water cycle in the urban areas. Both diffuse and point-source pollution is the main cause of the water bodies' quality deterioration, referring to both industrial wastewater and municipal sewage (WRC, 2007).

Municipal sewage is a mixture of domestic wastewater, non or partially treated industrial wastewater and rain or storm water. On a global scale, sewage represents the main point-source of water pollution (Gijzen, 2002). At present, about 2.4 billion people in the world have no access to improved sanitation facilities, mainly concentrated in developing regions (United Nations, 2015). In the majority of the developing countries wastewater is directly discharged into the sea, lakes, wetlands, lands and rivers without receiving any treatment. This situation acquires a particular connotation in urban areas where exists the possibility of mixing the sewage with untreated industrial wastewater, which represents an important pollution load to the environment (Björklund *et al*, 2009).

Human societies are mainly responsible for excessively discharging nitrogen compounds into freshwater and saltwater ecosystems through diverse sources and pathways. The transport of nitrogen compounds impairs the water cycle as depicted in Figure 1.1. The primary pathways of nitrogen pollution are: air, surface water or groundwater. The anthropogenic sources of nitrogen contamination include sewage, industries, septic tanks, urban storm water runoff, agriculture, livestock operations, aquaculture and fossil fuel combustion (Selman and Greenhalgh, 2009).

Despite the fact that diffuse sources control the inputs of nitrogen in most areas of the world, human excreta (urine) is the predominant source of nitrogen contamination in urban regions (Selman and Greenhalgh, 2009). The world population living in these areas is 54% and by 2050 is expected to be 66%; on the other hand, rural population is projected to decline 6% by 2050 (United Nations, 2014). Therefore, the current and future control of nitrogen pollution from sewage in urban areas is of eminent importance.

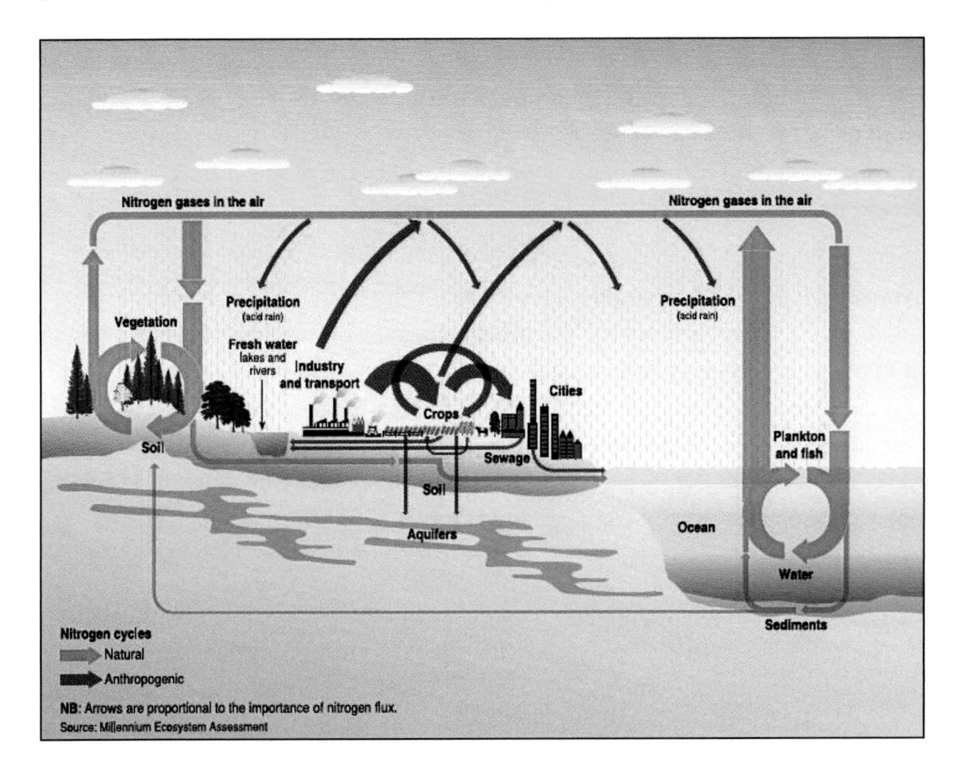

Figure 1.1. Interaction of the Nitrogen Cycle with the Water Cycle (Source: Millennium Ecosystem Assessment Board, 2005).

1.1.2. Transport and fate of inorganic nitrogen

Rivers have a dynamic role in the transport of nitrogen to the coasts, mainly in the urban areas impaired by human sewage (Howarth and Marino, 2006; Selman and Greenhalgh, 2009). For instance, human sewage provides 12% of riverine nitrogen input in the United States, 25% in Western Europe, 33% in China and 68% in the Republic of Korea (Howarth *et al.*, 1996, 2002; NRC 2000; Bashkin *et al.*, 2002; Xing and Zhu 2002).

The disparity of nitrogen input is caused by the variations in availability of infrastructure for sewage treatment and applied technologies. Thus, more than 80% of sewage in developing countries is discharged untreated and in those cases where sewage is treated, usually the treatment is not directed to nutrient removal, i.e. nitrogen. In industrialized countries, large efforts have been made related to nutrient removal from sewage. For example, in the last 20 years Europe's Urban Wastewater Treatment Directive has provoked an increasing number of sewage treatment facilities performing nutrient removal (Björklund *et al.*, 2009). Some of the technologies utilized can remove up to 90% of the nitrogen from sewage (Howarth *et al.*, 2005). However, the current technological approach for nitrogen removal from sewage needs to reconsider its long-term effects on the urban watersheds.

Mayorga *et al.* (2010) have proposed a model of the nitrogen export from watersheds highlighting sewage conveyance and treatment, and the transport of nitrogen from land to rivers and finally to the coast (Figure 1.2). Nitrogen in sewage usually consists of: (i) dissolved inorganic nitrogen (DIN): ammonia nitrogen (typically 20-75 mg/L), nitrate nitrogen + nitrite nitrogen (0.1-0.5 mg/L) and (ii) organic nitrogen: dissolved organic nitrogen (DON) + particulate nitrogen (PN) reaching around 15-25 mg/L (Henze and Comeau, 2008; Seitzinger and Harrison, 2008).

Seitzinger *et al.* (2010) have applied the models of the system Global Nutrients Export from Watersheds 2 (NEWS 2) to study the global trends in nutrient export by rivers and the nitrogen inputs from sewage were included (Seitzinger and Harrison, 2008). The models have taken into account the net effect of several factors including nitrogen removal by sewage treatment.

The results from NEWS 2 demonstrate that between the years 1970 and 2000, the DIN load to the coast increased 30%. The DON load also rose about 5% for the same period. Similarly, the total nitrogen load (TN= DIN+DON+PN) exported by rivers for the year 2000 was calculated to be 43 Tera grams (Tg) of nitrogen while in 1970 this was estimated as 37 Tg of nitrogen.

According to the NEWS 2 system, the estimation of DIN input from treated sewage by year 2030 shows an increase, relative to year 2000, for all scenarios in: North America (0.3 Tg-N/year), South America (0.1-0.2 Tg-N/year), Africa (0.1-0.2 Tg-N/year), Europe (0.3-0.4 Tg-N/year) and South Asia (0.4-0.5 Tg-N/year).

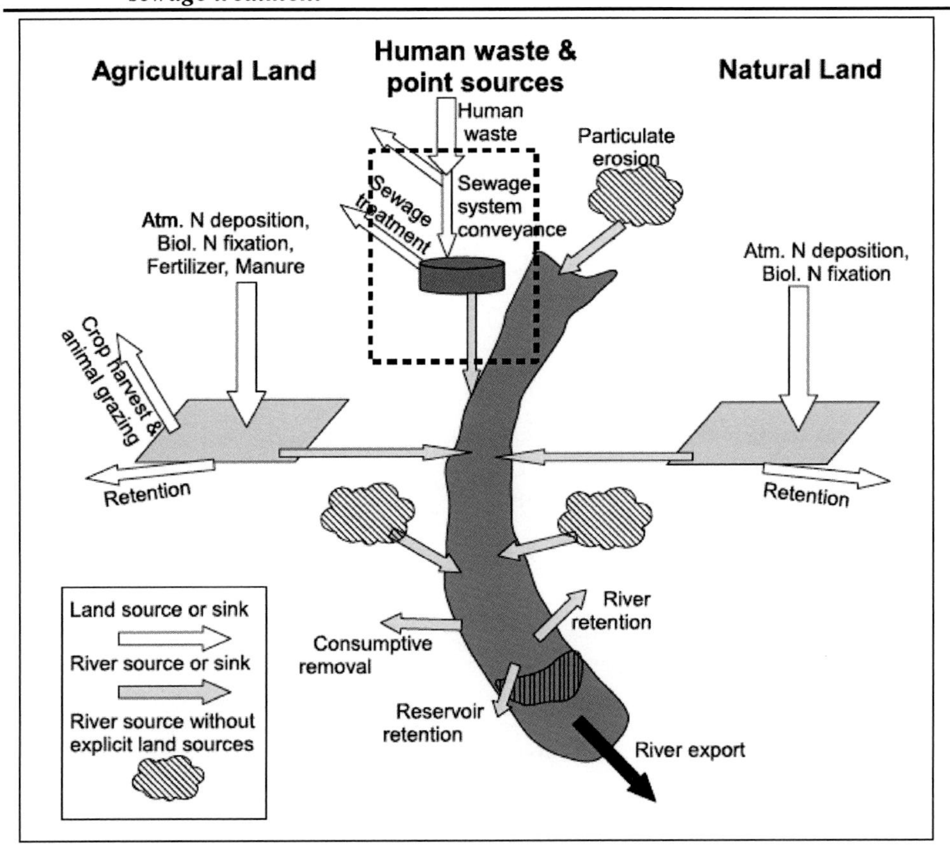

Figure 1.2. Nitrogen sources and sinks from watershed (land-surface), sources to rivers and export to the river mouth (Source: adapted from Mayorga *et al.*, 2010).

1.1.3. Impact of inorganic nitrogen

The nitrogen cycle is shown in Figure 1.3. Direct or indirect inputs of nitrogen have a considerable influence on the productivity of aquatic ecosystems (UNEP-GEMS/Water Programme, 2008). An aquatic ecosystem with a high productivity is defined as eutrophic. Some water bodies become eutrophic in a natural way, whereas others have become eutrophic because of sewage discharge. For this reason, countries of North America and the European Union have regulated the concentration of total nitrogen of sewage discharges. The allowed level of total nitrogen for discharges is set depending on sensitive areas, e.g. a range of 1.5-10 mg-TN/L (Oleszkiewicz and Barnard, 2006; Oenema, 2011). Nitrogen from sewage is one of the primary drivers of eutrophication (UNEP-GEMS/Water Programme, 2008). Björklund *et al.* (2009) have designated eutrophication as the most prevalent water quality problem worldwide. Already in 2005, the Board of Directors of the Millennium Ecosystem Assessment had identified nutrient pollution as one of the most outstanding environmental problems (Millennium Ecosystem Assessment Board, 2005). For instance, Figure 1.4 shows the initial indications of eutrophication in the second half of the past century and currently.

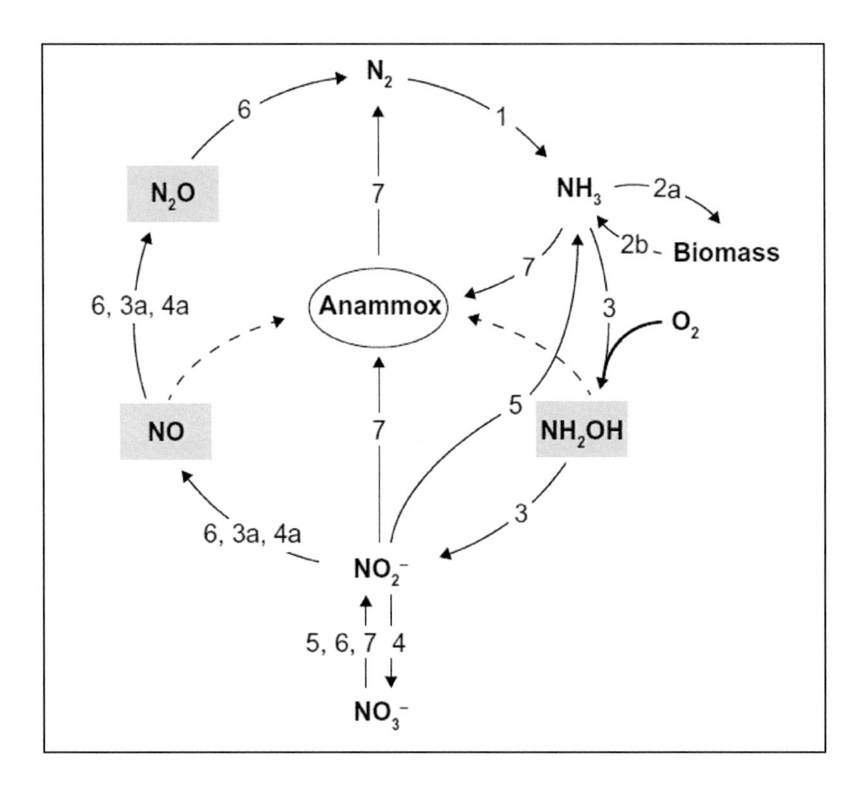

Figure 1.3. The nitrogen cycle: depicting the position of Anammox bacteria and partial oxidation and reduction processes: (1) nitrogen fixation; (2a) ammonium assimilation; (2b) ammonification/mineralization; (3) aerobic ammonium oxidation; (4) nitrite oxidation; (3a) and (4a) anaerobic nitrification-denitrification by ammonium and nitrite oxidizing nitrifiers, respectively; (5) dissimilatory nitrite reduction to ammonium (DNRA); (6) anaerobic denitrification; (7) anammox (Source: adapted from Kartal *et al.*, 2012).

Nitrogen is the primary cause of eutrophication in coastal ecosystems. Nowadays, about 500 coastal areas have been identified as suffering from eutrophication (Selman and Greenhalgh, 2009). The most severe consequence of eutrophication is hypoxia, i.e. a dissolved oxygen concentration less than 2 mg/L (Selman *et al.*, 2008).

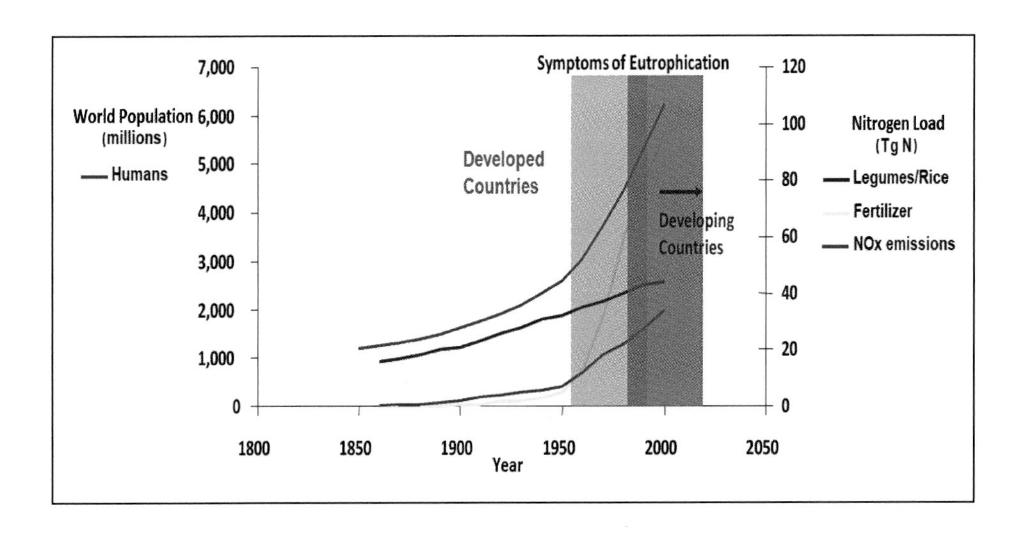

Figure 1.4. Period in which the symptoms of eutrophication began in developed countries and how the symptoms are shifted in recent years in developing countries (Source: modified from Rabalais *et al.*, 2010).

Hypoxia has increased over the past 50 years as a consequence of nitrogen input from urban sewage (Selman *et al.*, 2008; Selman and Greenhalgh, 2009; Rabalais *et al.*, 2010). The current state of hypoxic areas worldwide is described in Figure 1.5. In order to assess the future tendencies of eutrophication caused by nitrogen, an Indicator of Coastal Eutrophication Potential (ICEP) has been proposed by Billen and Garnier (2007). The positive values of ICEP indicate an excess of nitrogen leading to blooms of harmful species. The evaluation of this indicator up to 2050 shows an increment of positive ICEP values (expressed by the land area draining into the world's oceans) in all scenarios, i.e. a growing tendency to coastal eutrophication of world oceans, except for Arctic Ocean (Figure 1.6) (Garnier *et al.*, 2010).

1.2. Biological removal of organic carbon and dissolved inorganic nitrogen from sewage

Biotechnological processes have been widely adopted as the preferred treatment methods towards reducing the environmental pollution and the stress caused by sewage on natural resources because of less emission of chemical wastes, more cost-effective process applications, etc. (Khanal, 2008). Biological treatment processes may be classified according to their main electron acceptor (aerobic, anoxic, anaerobic), growth (suspended, attached) and combinations thereof.

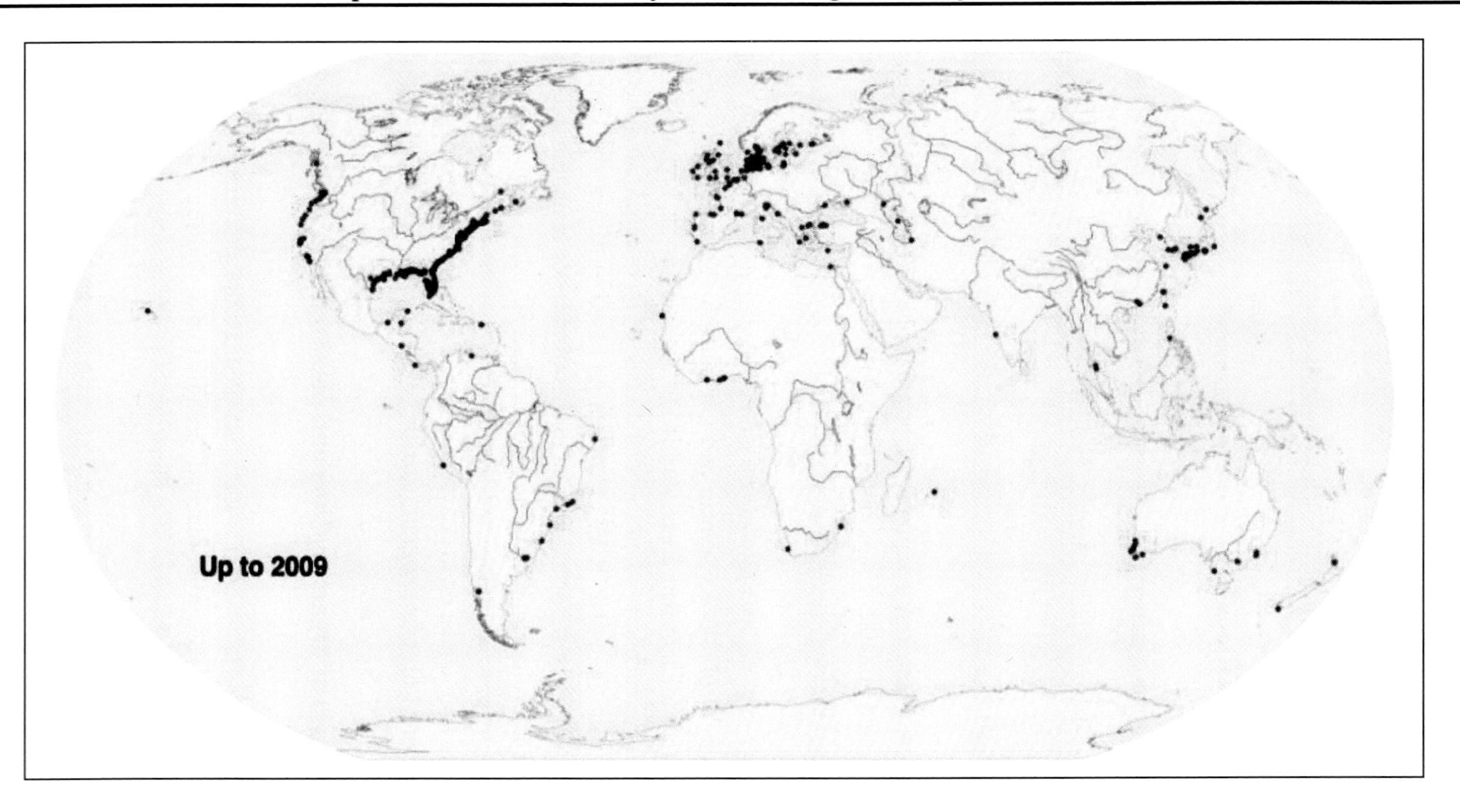

Figure 1.5. Global coastal hypoxia (occurrence of marine dead zones). Each red dot represents a documented case related to human activities (Source: Rabalais *et al.*, 2010).

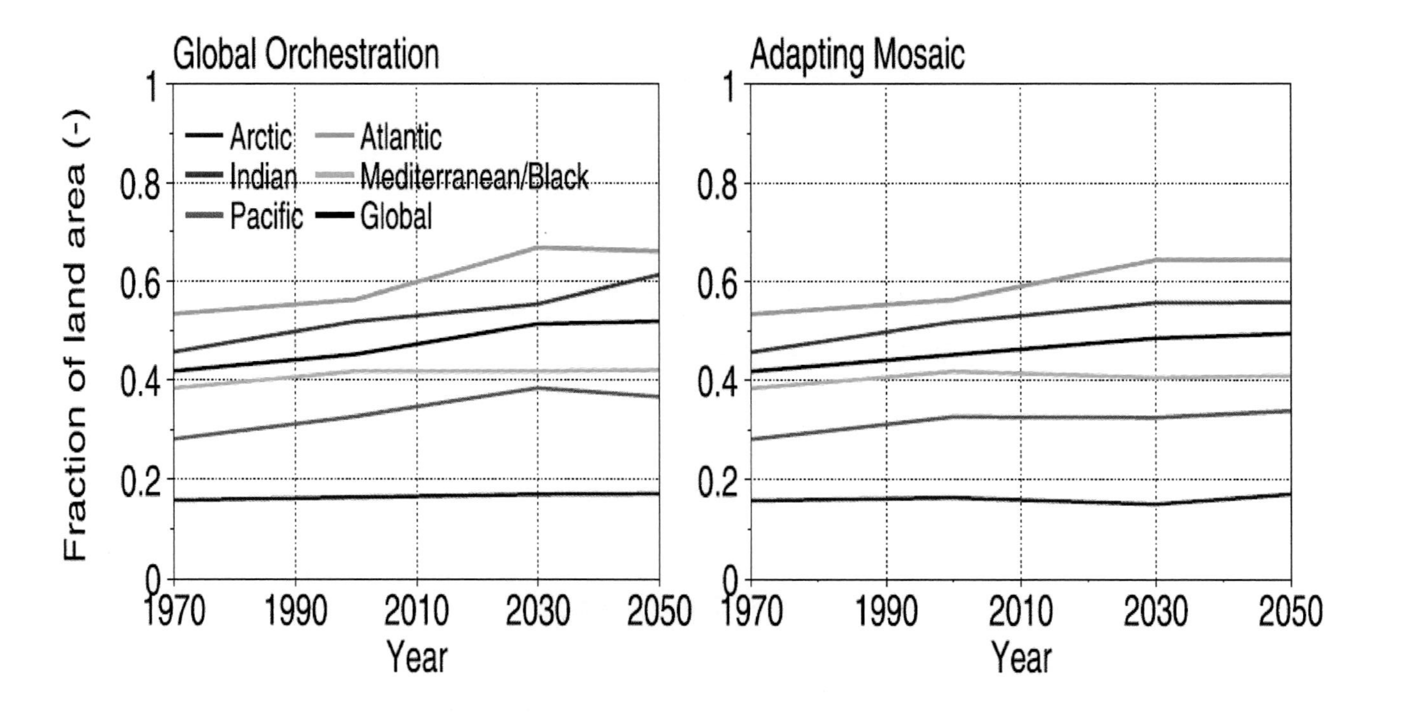

Figure 1.6. Fraction of land area with river basins draining into the world's oceans with ICEP > 0 for 2000-2050 for Global Orchestration (reactive approach) and Adapting Mosaic (proactive approach) scenarios (Source: Garnier *et al.*, 2010).

1.2.1. Sewage treatment by technology based on aerobic-anoxic stages

1.2.1.1. Mainstream sewage treatment through aerobic-anoxic stages

Regarding organic carbon and nitrogen removal, conventional activated sludge is the most widely applied aerobic wastewater treatment technology. This treatment process promotes the growth of flocculent biomass by supplying oxygen to the wastewater through aeration. The flocculent biomass contains microorganisms that metabolize the organic matter contained in wastewater (Tchobanoglous *et al.*, 2003). Overall, activated sludge has been successfully applied to obtain organic carbon removal with efficiencies in the range of 75-90% (Ekama and Wentzel, 2008a).

Anthropogenic nitrogen in municipal sewage originates from human excreta and urine, which increases with an increased consumption of proteins as part of the human diet (Mulder, 2003). The worldwide protein consumption is steadily increasing, e.g. the average worldwide per capita meat consumption has risen 87% (FAO, 2009). Therefore, it is of major importance to achieve a satisfactory nitrogen removal in the sewage mainstream treatment line. In terms of nitrogen control, the conventional activated sludge process can be extended with anoxic and low loaded aerobic stages with the aim to induce its removal by autotrophic nitrification (oxidation of ammonium to nitrate) and denitrification (reduction of nitrate to dinitrogen gas).

Several configurations for nitrogen removal in the main treatment line of the activated sludge systems have been developed based on anoxic and aerobic reactors as well as mixed liquor and sludge recycling lines, e.g. Wuhrmann, modified Ludzack-Ettinger and the 4 stage Bardenpho systems (Ekama and Wentzel, 2008b). Nevertheless, these nitrogen removal configurations are characterized by high energy consumption, whereas sufficient organic matter should be available as electron donor for denitrification. Applying internal recirculation flows or adding an external organic compound, e.g. methanol, to the system, can achieve the latter. In addition, proper operation of such nutrient removal activated sludge plant requires the implementation of excess sludge treatment and disposal.

Dutch researchers have developed a new mainstream aerobic system based on granular biomass. The microbial population is distributed in the aerobic granules in such way that key biological conversions can take place simultaneously, e.g. nitrification in the outer layer and denitrification in the inner region of the granules (de Kreuk, 2006). The first full-scale treatment plant based on this technology was opened on 2012 in the Netherlands with a capacity of 59,000 population equivalent (PE). This plant treats sewage that also contains wastes from slaughterhouses. A Dutch company, who denominated the system 'Nereda', commercializes the aerobic granular sludge system (Giesen *et al.*, 2013). Compared to the conventional activated sludge process, the Nereda process uses less energy and chemicals, has a lower footprint and less costs (Giesen *et al.*, 2013). Nowadays, 12 full-scale treatment plants are operating based on aerobic granular biomass (Pool, 2015).

1.2.1.2. Side-stream sewage treatment through aerobic-anoxic stages

The operation of activated sludge plants involves the treatment of internal process flows. For instance, sludge reject water coming from sludge handling facilities, i.e. side-stream treatment lines. This so-called 'sludge reject water' is usually diverted to the mainstream treatment process. Sludge reject water has a relatively higher temperature (20-35°C) and higher nitrogen concentration, i.e. ≈ 1000 mg-N/L (Lackner *et al.*, 2014). Because of this high nitrogen concentration, sludge reject water can contribute with 10-30% of the total nitrogen load (van Loosdrecht, 2008). In 1998, Hellinga *et al.* developed in the Netherlands the Single reactor High activity Ammonia Removal over Nitrite (SHARON) process, a biotechnological alternative to accomplish nitrite based nitrogen removal in side streams. Instead of conventional ammonium removal through nitrification and denitrification, the SHARON reactor accomplishes complete nitritation (equation 1.1) or partial nitritation (equation 1.2), i.e. ammonium oxidation to nitrite followed by heterotrophic nitrite reduction to dinitrogen gas.

$$2\,NH_4^+ + 3\,O_2 \rightarrow 2\,NO_2^- + 4\,H^+ + 2\,H_2O \qquad \textbf{(1.1)}$$

$$NH_4^+ + HCO_3^- + 0.75\,O_2 \rightarrow 0.5\,NH_4^+ + 0.5\,NO_2^- + CO_2 + 1.5\,H_2O \qquad \textbf{(1.2)}$$

Thus, the SHARON process can be carried out using two tanks configuration (separate aeration-nitrite reduction) or one tank system (sequential aeration-nitrite reduction). In general, the SHARON system represents substantial savings for nitrogen removal (25% less aeration and 40% less carbon source) and has a low sludge and carbon dioxide (CO_2) production, a reduction of 40% and 20%, respectively, with regard to conventional activated sludge (van Loosdrecht, 2008). The discovery by serendipity of anoxic ammonium oxidation (Anammox) bacteria opened the possibility for biological autotrophic nitrogen removal through the conversion of ammonium to dinitrogen gas under anoxic conditions with no organic carbon requirements (Mulder *et al.*, 1995; van de Graaf *et al.*, 1995; Kuenen, 2008).

Anammox bacteria employ nitrite (NO_2^-) as electron acceptor for ammonium oxidation and as electron donor for the reduction of carbon dioxide (CO_2), which is the source of carbon for autotrophic biomass production. The stoichiometry of Anammox bacteria has been described using either granulated (equation 1.3) and suspended cell (equation 1.4) Anammox cultures by Strous *et al.* (1998) and Lotti *et al.* (2014a), respectively:

$$NH_4^+ + 1.32NO_2^- + 0.066HCO_3^- + 0.13H^+ \rightarrow 1.02N_2 + 0.066CH_{1.8}O_{0.5}N_{0.2} + 0.26NO_3^- + 2.03H_2O \quad \textbf{(1.3)}$$

$$NH_4^+ + 1.146NO_2^- + 0.071HCO_3^- + 0.057H^+ \rightarrow 0.986N_2 + 0.071CH_{1.74}O_{0.31}N_{0.20} + 0.161NO_3^- + 2.002H_2O \quad \textbf{(1.4)}$$

Anammox bacteria belong to the *Planctomycetes* phylum. Table 1.1 depicts the genera and species proposed currently. Strous *et al.* (1997) recognized the potential of the Anammox process for ammonium removal from effluents of sludge digesters in the sewage treatment plants. Nowadays, this process has been successfully applied under such conditions (van der Star *et al.*, 2007; Wett, 2007; Siegrist *et al.*, 2008). For example, DIN removal is achieved by coupling SHARON and Anammox reactors (a two stage system). The first reactor is aerated and nitrite is produced by aerobic autotrophic ammonium oxidizing organisms (AOO) under an oxygen limitation regime, i.e. a partial SHARON process where about 50% of ammonium is oxidized, whereas in the second reactor the Anammox process takes place under anoxic conditions. This configuration can be found in the bibliography by the acronym SHARON-Anammox (van Dongen *et al.*, 2001) or other names (Trela *et al.*, 2004; Wyffels *et al.*, 2004). Table 1.2 depicts the benefits of the SHARON-Anammox process compared to the conventional nitrification-denitrification system.

A second alternative for DIN removal from the reject water of the sludge treatment facilities is the one stage partial nitritation-Anammox system. In this scheme, the biomass growth is promoted in different forms of aggregates, e.g. compact biofilms (fixed biomass), sludge granules and suspended biomass (Hippen *et al.*, 1997; Third *et al.*, 2001; Joss *et al.*, 2009). Two differentiated zones can be distinguished in these sludge aggregates: an external nitritation zone where ammonium is partially oxidized to nitrite by a dominant population of AOO and an internal zone dominated by Anammox bacteria. In the internal region the nitrite generated from the external layer and the remaining ammonium are mainly transformed to dinitrogen gas (Figure 1.7).

Table 1.1. Anammox genera and species provisionally proposed.

Anammox Bacteria			
Genera	**Proposed Species**	**Origin**	**Reference**
Ca. Brocadia	*Ca. Brocadia anammoxidans*[a]	Denitrifying fluidized bed reactor	Strous *et al.*, 1999
	Ca. Brocadia fulgida[a]	Sequencing batch reactor	Kartal *et al.*, 2008
	Ca. Brocadia sinica	Up-flow fixed-bed anammox biofilm reactor	Oshiki *et al.*, 2011
	Ca. Brocadia caroliniensis	Anammox bioreactor	Rothrock *et al.*, 2011
Ca. Kuenenia	*Ca. Kuenenia stuttgartiensis*[a]	Trickling filter	Schmid *et al.*, 2000
Ca. Scalindua	*Ca. Scalindua sorokinii*	Black Sea	Kuypers *et al.*, 2003
	Ca. Scalindua brodae	Rotating biological contactor	Schmid *et al.*, 2003
	Ca. Scalindua wagneri	Rotating biological contactor	Schmid *et al.*, 2003
	Ca. Scalindua arabica	Arabian Sea	Woebken *et al.*, 2008
	Ca. Scalindua sinooilfield	Petroleum reservoir	Li *et al.*, 2010
	Ca. Scalindua zhenghei	South China Sea	Hong *et al.*, 2011
	Ca. Scalindua richardsii	Black Sea	Fuchsman et al., 2012
	Ca. Scalindua pacifica	Bohai Sea	Dang *et al.*, 2013
	Ca. Scalindua profunda	Swedish fjord	van de Vossenberg *et al.*, 2013
Ca. Anammoxoglobus	*Ca. Anammoxoglobus propionicus*[a]	Sequencing batch reactor	Kartal *et al.*, 2007
	Ca. Anammoxoglobus sulfate	Non-woven rotating biological contactor reactor	Liu *et al.*, 2008
Ca. Jettenia	*Ca. Jettenia asiatica*	Up-flow granular sludge anammox reactor	Quan *et al.*, 2008
	Ca. Jettenia moscovienalis	Sludge digester of a sewage treatment plant	Nikolaev *et al.*, 2015
	Ca. Jettenia caeni[b]	Membrane bioreactor	Ali et al., 2015
Ca. Anammoximicrobium		Enriched culture from Moscow River sample	Khramenkov *et al.*, 2013

[a] These species couple the oxidation of formate, acetate and propionate to CO_2 with the transformation of nitrate or nitrite to dinitrogen gas; the highest specific oxidation rates for *Ca. B. fulgida* and *Ca. A. propionicus* correspond to acetate and propionate, respectively (Kartal *et al.*, 2008).
[b] It is able to reduce nitrate with the oxidation of acetate to CO_2 and the conversion of nitrate to dinitrogen gas (Ali *et al.*, 2015).

Table 1.2. Comparison between SHARON-Anammox process and conventional nitrification-denitrification system for nitrogen removal (Source: van Loosdrecht, 2008).

Item	Unit	Conventional treatment	SHARON®/Anammox
Power	kWh/kg-N	2.8	1.0
Methanol	kg/kg-N	3.0	0
Sludge Production	kg-VSS/kg-N	0.5-1.0	0.1
CO_2 emission	kg/kg-N	> 4.7	0.7
Total costs[1]	€/kg-N	3.0-5.0	1.0-2.0

[1]Total costs include both operational costs and capital charge.

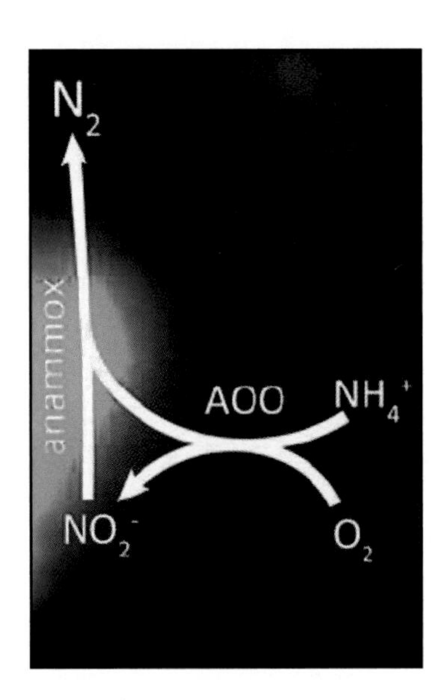

Figure 1.7. Cross-section of a granule from one stage Anammox system: the AOO and Anammox bacteria are depicted in red and green, respectively by fluorescence in situ hybridization. Ammonium conversion to dinitrogen gas is shown by the overlaid reactions. (Source: adapted from Hu *et al.*, 2013).

Completely Autotrophic Nitrogen-removal Over Nitrite (CANON) is one example of the names used for the designation of this system (Third *et al.*, 2001). In a worldwide survey of 100 full-scale Anammox treatment plants utilized for DIN removal in landfill leachate, industrial wastewater and sewage, Lackner *et al.* (2014) found that the major use of Anammox technology is in the side-stream lines of sewage treatment plants, representing the 75% of the surveyed plants. Diverse types of reactors and configurations are utilized in these treatment plants, e.g. sequencing batch reactors (SBR), moving bed biofilm reactors (MBBR), SHARON/ANAMMOX®, integrated fixed film activated sludge (IFAS), etc. A detailed list of full-scale Anammox treatment plants can be found in Lackner *et al.* (2014). Figure 1.8 shows their geographical distribution (Ali and Okabe, 2015).

1.2.1.3. Advancements on the treatment of the mainstream sewage by using Anammox technology

With the objective of optimizing their energy consumption efficiency, literature shows a strong interest in the implementation of DIN removal using Anammox based-technology in the mainstream of sewage treatment plants, e.g. Gao *et al.* (2014), Hendrickx *et al.* (2014), Lotti *et al.* (2014b), Morales *et al.* (2015) and Schaubroeck *et al.* (2015). The strategy consists on decoupling the removal of chemical oxygen demand (COD) and nitrogenous oxygen demand (NOD) in such way that the carbon sources would be utilized for energy generation while the energy requirements for nitrogen removal would be diminished. By following this approach, an energy-neutral or even energy-positive sewage treatment process can be achieved. A summary of the alternatives proposed by some research groups to obtain this neutral or positive energy balance can be found in Gao *et al.* (2014) and Morales *et al.* (2015). Basically, the alternatives proposed can be categorized in (i) a variation of aerobic sewage treatment-anaerobic sludge digestion technology in combination with Anammox reactors and (ii) the coupling of anaerobic and Anammox reactors.

There are some cases where Anammox activity was achieved coincidentally in the mainstream. Subsequently, various researchers tried to develop and implement a full-scale mainstream Anammox process. Recently, Cao *et al.* (2013) have reported a significant and spontaneous Anammox activity in the mainstream of the largest full-scale activated sludge system in Singapore treating 800.000 m^3 sewage/d. This sewage treatment facility has a step-feed activated sludge process with five aerobic/anoxic zones.

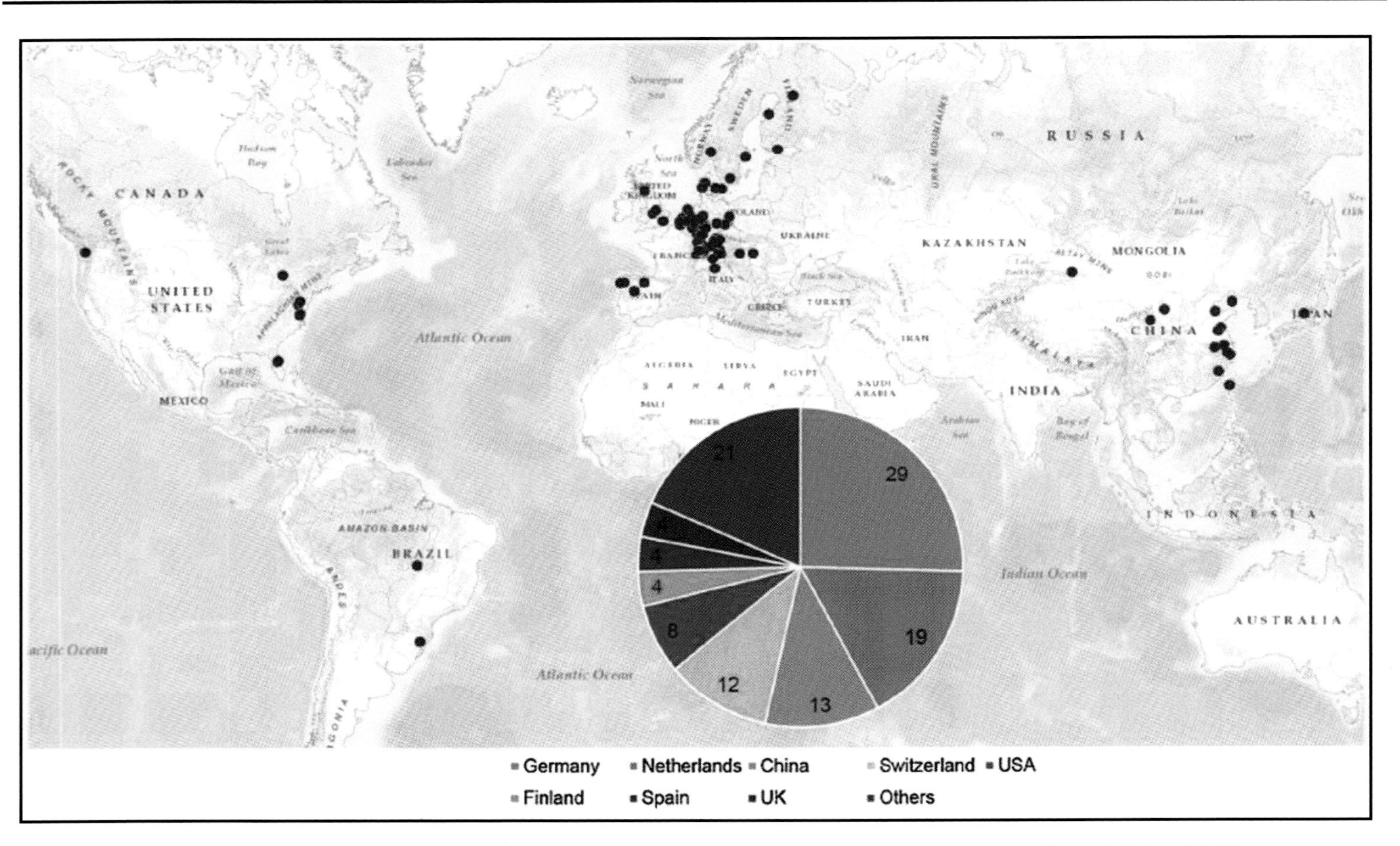

Figure 1.8. Distribution of full-scale Anammox treatment plants worldwide (Source: Ali and Okabe, 2015).

Its configuration and the operational conditions applied may be the driving forces that caused an unexpected and stable partial nitritation in which 64.6% of the ammonium was oxidized to nitrite in the aerobic zones, i.e. temperature of 28-32°C, the alternation of aerobic and anoxic regions, and a solids retention time (SRT) of only 2.5 days. Although the average Chemical Oxygen Demand (COD) to ammonium nitrogen (NH_4^+-N) ratio at the inlet points of the anoxic regions was 7.3 g-COD/g-NH_4^+-N, the ammonium removed in the anoxic zones was 50% of the influent ammonium concentration and the total contribution of the autotrophic nitrogen removal was 100% and 75% of ammonium and total nitrogen, respectively.

More recently, Professor Mark van Loosdrecht, from TU Delft University, has proposed the coupling of the Nereda system with the Anammox process (Pool, 2015); this represents a possibility for using this treatment scheme in the mainstream for DIN removal from sewage (Winkler *et al.*, 2012). Furthermore, Isanta *et al.* (2015) have achieved a stable partial nitritation in an aerobic granular sludge reactor using low-strength wastewater and Cydzik-Kwiatkowska and Wojnowska-Baryla (2015) have demonstrated the viability of the development of Anammox microorganisms in aerobic granules, finding that the hydraulic retention time (HRT) and the oxic conditions, i.e. anoxic/aerated mode of operation in an SBR, determine the growth of Anammox bacteria in the aerobic granules.

Lotti *et al.* (2015) have suggested to modify the actual treatment scheme of the Dokhaven treatment plant of Rotterdam, the Netherlands. Its original configuration consists of an A-B system, i.e. the first reactor (A-stage) has a high load and the biochemical oxygen demand (BOD) is transformed in biomass. In the second reactor (B-stage), the removal of the remaining BOD and nitrification of ammonium is achieved. This plant has also a treatment for the excess of sludge in a side stream and includes anaerobic digestion and nitrogen removal via the SHARON-ANAMMOX® process. The proposal of the new configuration consists of the replacement of the B-stage by a one stage partial nitritation-Anammox reactor; the treatment scheme for the side stream remains the same. The recommendation is supported by the removal attained in the pilot plant, confirms the potentials of the suggested configuration (Lotti *et al.*, 2015).

When using aerobic sewage treatment including anaerobic sludge digestion in combination with Anammox reactors for DIN removal in the mainstream the constraints related to the aerobic stage needs to be taken into account: (i) investment, operational and maintenance costs, (ii) high energy demand for aeration, (iii) the generation of an excess of sludge and its treatment, etc. These aspects would be avoided or diminished by the use of an anaerobic treatment - Anammox configuration for the mainstream.

1.2.2. Mainstream sewage treatment by means of anaerobic technology

Under the total absence of oxygen in anaerobic wastewater treatment, diverse groups of microorganisms convert the biodegradable organic matter from sewage in biomass and biogas, mainly consisting of methane (CH_4), a source of energy, and carbon dioxide (CO_2). Anaerobic sewage treatment also allows the stabilization of the sludge retained in the system, which improves the dewatering characteristics of the sludge. Anaerobic sewage treatment is characterized by a positive energy balance because it avoids the use of electricity as energy source for treatment (usually produced by fossil fuels) whereas energy-rich CH_4 is generated. In addition, anaerobic wastewater technology has other advantages such as (i) a lower sludge production, (ii) lower footprint, (iii) does not require highly skilled labor force, (iv) relatively high treatment efficiencies, etc. (van Lier *et al.*, 2008). Several anaerobic systems and configurations have been developed through the years and some of them have been applied for sewage treatment. For instance, the first full-scale anaerobic sand filter used for sewage treatment was introduced in an experimental station of Massachusetts, USA, in 1887. This anaerobic attached growth system was not very popular because of odor problems and the limited knowledge of the microbiology and the sanitary engineering aspects inherent to the anaerobic process. However, in the 1960's an enhanced version of the anaerobic filter was established in USA (McCarty 1964, 2001).

At present, high rate anaerobic reactor systems are being used, in which the solids retention time and the hydraulic retention time are uncoupled. This feature allows that the maximum permissible load is ruled by the quantity of anaerobic bacteria in contact with the wastewater. A high retention of viable microorganisms can be obtained through (i) biofilm attachment to fixed or non fixed carriers, (ii) microbial auto immobilization, e.g. flocs or granules and (iii) separation techniques, e.g. sludge settling and membrane filtration.

The full-scale application of high rate anaerobic systems for the treatment of industrial effluents is well documented. The variety of industries includes: pulp and paper industry, beverage, agro-food industry, alcohol distillery, pharmaceutical, chemical, etc. (van Lier *et al.*, 2008). The three most popular high rate anaerobic configurations utilized for industrial wastewater treatment are: (i) the expanded granular sludge bed (EGSB), 22%; (ii) the internal circulation (IC) reactor (a variation of the EGSB with biogas-driven hydrodynamics), 33% and (iii) the upflow anaerobic sludge blanket (UASB) reactor, 34% (van Lier, 2007).

EGSB reactors have important advantages compared to UASB reactors, e.g. provide an enhanced contact between the granular biomass and the wastewater which reduce the mass transfer limitations, an improved hydraulic mixing, a high performance during the treatment of low-strength and high strength wastewater, very small footprint, etc. (van Lier *et al.*, 2008). For successfully operating EGSB reactor the presence of high quality granular sludge is indispensable (van Lier *et al.*, 2015). With regard to full-scale anaerobic sewage treatment, the UASB reactor is commonly applied, owing to its simplicity and low investment and operational costs (Chernicharo *et al.*, 2015; Lettinga and Hulshoff Pol, 1991).

1.2.2.1. Mainstream sewage treatment by upflow anaerobic sludge blanket (UASB) reactor

At the beginning of the 1970's, the UASB was developed in the Netherlands (Lettinga, 1980). Nowadays, the UASB reactor is the most popular configuration of anaerobic sludge bed reactors for municipal sewage treatment (Figure 1.9).

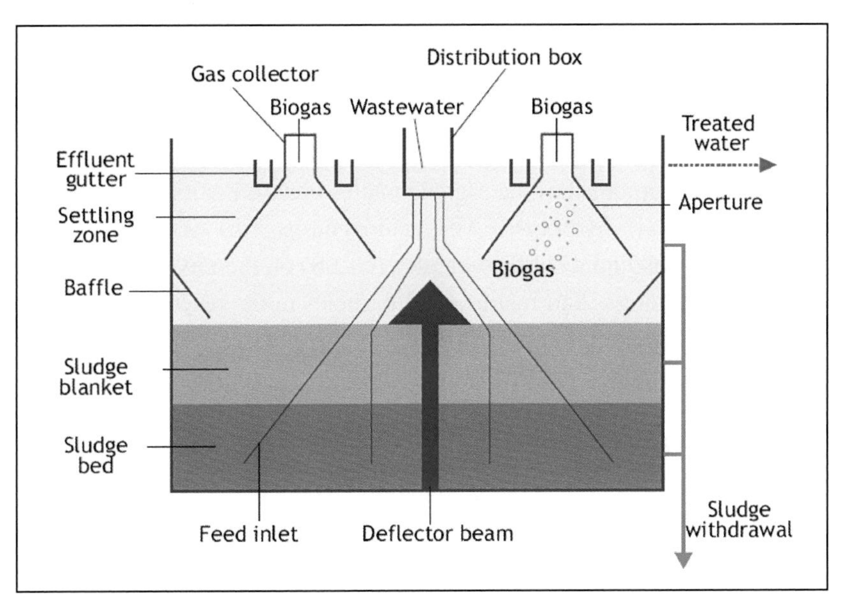

Figure 1.9. Scheme of a UASB reactor for sewage treatment. Most relevant design characteristics are depicted (Source: van Lier *et al.*, 2008).

The UASB reactor is a system where three phases coexist: the liquid (sewage), the solids (anaerobic biomass) and the gas (produced biogas). The characteristics of the sewage, as well as the environmental and operational factors influence the treatment of sewage in UASB reactors. Sewage is a very complex wastewater with a low COD concentration (usually less than 1000 mg-COD/L), a high fraction of COD as suspended solids (approximately 50-65%), fatty compounds, proteins, detergents, etc. and temperatures less than < 30°C (Chernicharo *et al.*, 2015; Foresti, 2002; van Lier *et al.*, 2008). Also, the availability of drinking water impacts the characteristics of sewage, resulting in more concentrated sewage when the potable water consumption is limited. For instance, in Palestine the concentrations of total COD, total Kjeldahl nitrogen (TKN) and ammonium nitrogen are in the range of 1092-3812 mg-COD/L, 54-122 TKN/L and 40-89 mg-NH_4^+-N/L, respectively (Mahmoud *et al.*, 2003).

The mixing in UASB reactors takes place by the upward influent flow and the turbulence brought forward by the uprising biogas. Inside the UASB, the incoming sewage flows in ascendant direction, expanding the anaerobic sludge bed. The sludge bed consists of biomass, which grows forming dense flocs and, if conditions allows, in granules. These biological aggregates consist of associations of diverse groups of bacteria: hydrolytic and fermentative bacteria, acetogenic bacteria, homo-acetogenic bacteria, hydrogenotrophic methanogens and acetoclastic methanogens.

The anaerobic sewage treatment takes place in the sludge bed through suspended solids accumulation and organic matter conversion into biogas and biomass. The anaerobic biomass accomplishes the organic matter transformation to biogas using the following pathway: (i) hydrolysis, (ii) acidogenesis, (iii) acetogenesis and (iv) methanogenesis. A gas/liquid/solids separator (GLSS) on the top of the reactor allows the collection of the biogas and retains solid particles in the reactor. The reduction of organic matter in sewage through UASB reactors, expressed as BOD and COD, has been reported in the range of 70-85% and 65-80%, respectively (Chernicharo *et al.*, 2015; van Lier *et al.*, 2008); although, lower treatment values are commonly attained in India attributable to poor system designs and poor operation and management (van Lier *et al.*, 2010).

The tropical and subtropical climate zone offers optimal temperatures (usually > 20°C) for sewage treatment by UASB reactors (Chernicharo *et al.*, 2015; Schellinkhout and Osorio, 1994; van Haandel and Lettinga, 1994). During the last 25 years, the full-scale anaerobic treatment of sewage by means of UASB technology has grown and reached a mature status mainly in the developing countries; in some cases the population equivalent capacity attained is up to one million inhabitants (Chernicharo *et al.*, 2015; Chernicharo, 2006).

In despite of the needs to improve certain design and operation aspects of UASB reactors e.g. scum accumulation, odor emission, methane losses, energy recovery and effluent post-treatment (Chernicharo *et al.*, 2012, 2015), currently the technical solutions for these drawbacks are available or under development. For example, the removal of methane in effluents from UASB reactors (Hatamoto *et al.*, 2010; Bandara *et al.*, 2011) or the control of scum (Rosa *et al.*, 2012). Therefore, the sewage treatment by UASB reactors is nowadays a consolidated technology having a high capability to reduce organic pollution loads from sewage (Noyola *et al.*, 2012). There are successful examples of full-scale UASB applications for sewage treatment in countries of Latin America, Asia, Middle East and Africa (van Lier *et al.*, 2010; Chernicharo *et al.*, 2015), e.g. Colombia (van Haandel and Lettinga, 1994; Giraldo *et al.*, 2006), India (Khan *et al.*, 2014), Brazil (Chernicharo *et al.*, 2009), Mexico (Monroy *et al.*, 2000), Guatemala (Sánchez de León, 2001), Egypt (van Lier *et al.*, 2008), United Arab Emirates (Heffernan *et al.*, 2011), Ghana (de Mes *et al.*, 2004; Awuah and Abrokwa, 2008).

Similar to other anaerobic wastewater treatment technologies, the effluent produced by the UASB reactor generally requires an additional treatment before its final disposal in order to meet the criteria for its re-use or discharge, e.g. additional removal of organic matter, removal of pathogens and nutrients like nitrogen, mainly as NH_4^+-N, etc. (van Lier *et al.*, 2001; Chernicharo, 2006, Chernicharo *et al.*, 2015).

The options for post-treatment of UASB effluents comprise both mechanized treatment and natural-based technologies. Mechanized treatment technologies include systems based on physicochemical and biological processes. Examples of nitrogen removal post-treatment systems by biological processes are: activated sludge systems, trickling filters, rotating biological contactors, membrane technology, etc. Since the organic load in UASB effluents will be too low for the biological removal of nitrogen via heterotrophic growth (like in aerobic biological systems), post nitrification-denitrification requires the need for an external carbon source in addition to energy supply for aeration, etc., which increases the investment, maintenance and operating cost of this alternative.

The post-removal of nitrogen through the use of natural based-technologies includes systems such as: maturation (polishing) ponds, constructed wetlands, over land flow and aquaculture systems (e.g. use of macrophytes) (von Sperling and Chernicharo, 2005; Kujawa-Roeleveld, 2011). The group of von Sperling and collaborators found a good performance of maturation ponds in series to treat the effluent of a full-scale UASB reactor used for sewage treatment (Dias *et al.*, 2014). Also, Ferreira da Costa *et al.* (2015) have reported the use of constructed wetlands as post treatment systems for full-scale UASB effluents. However, the application of natural based-technologies for post-treatment of the UASB effluents depends on the land availability and soil characteristics, which in many cases is limited or inadequate for large natural systems, mainly in the urban areas. In order to overcome these circumstances, the use of more compact systems has been recommended, like trickling filters (Chernicharo and Nascimento, 2001; Almeida *et al.*, 2009). Some research groups have tested trickling filters, as UASB post-treatment units, with different packing media, e.g. plastic rings, parts of corrugated plastic tubing, blast furnace slag, Rotopack (plastic-based packing media), Rotosponge (plastic-based packing media), etc. (Procópio Pontes and Chernicharo 2011; Almeida *et al.*, 2013; Vieira *et al.*, 2013).

Currently, various types of trickling filters are in full-scale operation, while others are under study and development. For instance, an experimental system (for 500 inhabitants) consists of a trickling filter and a lamella sedimentation chamber that are connected to the UASB reactor as one compact unit (Procópio Pontes and Chernicharo, 2011). Other examples include trickling filters with Rotopack or Rotosponge as reported by Almeida *et al.* (2013), or the open trickling filter packed with stones and designed by Vieira *et al.* (2013).

Examples of full-scale trickling filters includes: (i) Two UASB sewage treatment plants in Minas Gerais State-Brazil that use trickling filters with rock as packing media, one plant serving a Population Equivalent (PE) of 70,000 inhabitants and the other plant with a PE capacity for 1,000,000 inhabitants (Chernicharo *et al.*, 2009), (ii) A UASB sewage treatment plant in Agra-India; the trickling filter uses sponge media as packing material and is based on the knowledge gained from the research with the down flow hanging sponge (DHS) systems. In fact, the latter trickling filter is the first full-scale application of DHS technology (Harada, 2014). Although trickling filters may improve the quality of some characteristics of the UASB effluents, the removal of DIN in the trickling filters mentioned above is not enough to comply with the regulations for effluent discharge (Vieira *et al.*, 2013). Therefore, it is necessary to apply a post-treatment scheme or configuration that allows meeting the corresponding DIN discharge standards mainly in urban regions of developing countries.

1.2.3. Mainstream sewage treatment by UASB and Anammox reactors

The integration of anaerobic treatment and Anammox process was recognized by Jetten *et al.* (1999) as a promising alternative to maximize the capacity of both processes. Similarly, Gujer (2010) have suggested the post-treatment of the effluents from the anaerobic reactors through the SHARON-Anammox process as a significant option for sewage treatment, mainly in warm climates.

The use of Anammox technology for the removal of DIN from the UASB effluent has several environmental and engineering advantages: (i) a low biomass yield per mol of ammonium, resulting in low sludge production; (ii) no need for an external organic carbon source because ammonium itself is used as electron source for denitrification, whereas CO_2 is used as a source of organic carbon for the metabolic processes; (iii) the costs of energy, operation & maintenance for aeration are not as high as in activated sludge systems for nitrogen removal (Table 1.2), since the Anammox process only requires to oxidize about 50% of ammonium to nitrite; and, (iv) the biomass has the ability to form aggregates as biofilms or granules which allows the design of compact systems (van Loosdrecht, 2008; van der Star *et al.*, 2008; Kartal *et al.*, 2010).

The feasibility of autotrophic nitrogen removal from UASB effluents would be mainly influenced by: (i) the UASB effluent characteristics which could affect the microorganisms involved in the transformation of the nitrogen compounds; (ii) the growth of other groups of microorganisms promoted by the characteristics of the UASB effluent and treatment conditions; and (iii) the design and operational features of the Anammox system (one or two-stage).

In general UASB effluents from a sewage treatment plant have suitable characteristics for DIN removal by Anammox process. For instance, there is sufficient alkalinity (Vieira *et al.*, 2013) and the pH is in the appropriate range for AOO and Anammox bacteria. Biological removal of DIN through these microorganisms can be carried out in a wide range of pH, i.e. 7 to 8 (van Hulle *et al.*, 2010) and 6.7-9.0 (Kartal *et al.*, 2012), respectively. Also the ammonium concentration in the effluent of the UASB reactor, i.e. a range estimated from 20-100 mg-N/L, and the expected C/N ratio is ideal for AOO and Anammox bacteria.

During anaerobic treatment organic compounds are converted to methane whereas the ammonium concentration increases owing to the mineralization of N containing organics such as proteins (Sanz and Fdz-Polanco, 1990; Mahmoud *et al.*, 2003; Henze and Comeau, 2008). Moreover, the concentrations of nitrite and ammonium after the partial nitritation of the effluent will be below the values reported as inhibitory for the AOO and the Anammox bacteria (van Hulle *et al.*, 2010; Lotti *et al*, 2012). Furthermore, the temperature of UASB effluents in tropical and subtropical regions is frequently higher than 20°C. This temperature range is more advantageous for the growth of AOO rather than nitrite oxidizing organisms (NOO). Similarly, temperatures above 20°C are adequate for the Anammox process since Anammox bacteria associated with non-saline environments have a growth temperature in the range of 20-45°C (Kartal *et al.*, 2007; Quan *et al.*, 2008; Kartal *et al.*, 2012; Khramenkov *et al.*, 2013; Ali *et al.*, 2015; Nikolaev *et al.*, 2015). Nevertheless, Lotti *et al.* (2014b) have reported the adaptation of Anammox bacteria in a one-stage Anammox lab reactor at 10°C during 100 days of operation treating synthetic sewage.

The first attempt to couple a UASB reactor and the Anammox process (two-stages: SHARON reactor followed by two SBR Anammox reactors operated at 25°C and 35°C) was in a decentralized system, where autotrophic nitrogen removal was performed from black water (average influent concentration to the Anammox reactor: 408 mg-NH_4^+-N/L and 483 mg-NO_2^--N/L); the SHARON-Anammox system achieved a removal of total nitrogen of 85-89% from the UASB effluent (de Graaff *et al.*, 2011). Other attempts for the coupling of UASB and Anammox lab-scale reactors, but for mainstream sewage treatment, have been reported by Malamis *et al.*, 2013 (Anammox one-stage SBR) and Malovanyy *et al.*, 2015 (Anammox one-stage MBBR). Nevertheless, both systems experienced problems that limited the performance of the biological processes. For instance, the AOO biomass suffered a severe detachment and washout from the Anammox one-stage MBBR used by Malovanyy *et al.* (2015). Furthermore, the total DIN in the final effluent of both UASB-Anammox systems was above 20 mg-N/L and did not comply with the regulations for effluent discharge.

There are a number of challenges that need to be addressed and solved first in order to guarantee the successful coupling of the UASB and Anammox reactors. Some of these challenges are related to the water quality of the UASB effluents, while others relate to the characteristics of Anammox process, e.g. the effect of the sulfide content in the UASB effluent on the Anammox process, the required outcompeting of nitrite oxidizing organisms (NOO), etc.

The tolerance of Anammox bacteria to sulfide has been tested in short-term experiments and the values for the 50% of inhibitory concentration (IC_{50}) have been reported to be dependent of the Anammox species, DIN concentration, temperature, pH and the form of the biomass aggregate, i.e. granules, biofilms, etc. Thus, the results for the IC_{50} for sulfide in short-term tests are in the range of 10-540 μM (Dapena-Mora *et al.*, 2007; Carvajal-Arroyo *et al.*, 2013; Jin *et al.*, 2013; Russ *et al.*, 2014 and Ali *et al.*, 2015).

Regarding the long-term effects of sulfide inhibition on the Anammox bacteria, the literature only reports the research of Jin *et al.* (2013). The long-term assay was carried out in a UASB reactor inoculated with Anammox granules and operated at 35 ± 1°C, pH 8.33 ± 0.18, influent concentrations of NH_4^+-N and NO_2^--N in the range of 125-325 mg-N/L (molar ratio NO_2^--N/NH_4^+-N around 1) and influent sulfide-S concentrations of 8-40 mg-S^{2-}/L. The reactor was started up with a period of 16 days of initial operation without feeding of sulfide-S and followed by a 25 days period using an influent sulfide-S concentration of 32 mg-S^{2-}/L. At the end of this period the maximum nitrogen removal rate (NRR) diminished by 67%. The sulfide-S and nitrogen compounds concentrations in the influent were adjusted and after an acclimatization period of 115 days the Anammox bacteria were able to tolerate 32 mg-S^{2-}/L. The drop in maximum NRR was 9% compared to the 'blank operation' without the presence of sulfide-S. According to the authors, the sulfide had a negative effect that caused an inhibition of the biomass growth rate of Anammox bacteria at the sulfide-S concentrations tested. Nevertheless, the presence of bacteria *Thiobacillus denitrificans* in the Anammox reactor suggests a competing of electron donors. *Thiobacillus denitrificans* reduces nitrogen compounds to dinitrogen gas via sulfide oxidation to elemental sulfur (Mahmood *et al.*, 2007). This metabolic pathway would be related with the drop in the maximum NRR.

Concerning the suppression of NOO in order to guarantee a stable nitrogen removal by the AOO-Anammox consortium, several studies have addressed different strategies for achieving this goal. Besides of the application of high temperatures to promote the prevalence of AOO over NOO (Hellinga *et al.*, 1998), some of the strategies for NOO suppression include: (i) the addition of high concentrations of inorganic carbon, i.e. by means of the use of $NaHCO_3$ (Tokutomi *et al.*, 2010), (ii) bioaugmentation with AOO and/or Anammox bacteria (Bartrolí *et al.*, 2011; Wett *et al.*, 2013), (iii) real-time control strategy of the length of aerobic and anoxic phases via pH and ORP sensors (Claros *et al.*, 2012), (iv) alternating anoxic and aerobic conditions, i.e. intermediate aeration (Wett *et al.*, 2013; Ge *et al.*, 2014), and (v) control of aeration, aerobic SRT and COD input, based on direct measurement of NH_4^+, NO_2^- and NO_3^- (Regmi *et al.*, 2014).

Up to now, one of the most common applied approaches for the NOO suppression is the control of dissolved oxygen (DO) concentration through intermittent aeration at a DO concentration up to 1.5 mg/L (Wett *et al.*, 2013). Nowadays, some research groups have focused the investigation on NOO suppression using modeled-based evaluations for mainstream ammonium removal. For instance, Pérez *et al.*, (2014) performed a model-based study for pretreated sewage at 10°C for a one-stage Anammox process. These authors found that although nitrite consumption is done by Anammox bacteria, the DO competition between AOO and NOO is crucial for NOO suppression. Other example is the research of Al-Omari *et al.*, (2015) who evaluated the configuration of the process and the specific features of the strategies applied via two approaches: (i) AVN controller: a target ratio of 1 for ammonia *vs.* NOx (nitrate+nitrite) and (ii) ammonia-based control; the strategy of the AVN controller showed a more efficient NOO suppression during the long-term research performed in pilot plants.

1.3. Aim and scope of the research

There is an increasing need to cope with the nitrogen pollution caused by sewage, e.g. in the urban coastal areas. In order to recover these coasts impaired by hypoxia and preserve those not affected by this condition, regulatory measures have been adopted through the promulgation of stricter environmental regulations for nitrogen discharges from sewage.

In response to this scenario and taking into consideration the worldwide financial crisis, it is of vital importance the development of robust, efficient, environmentally-friendly, low operational and maintenance requirements sewage treatment systems for DIN removal, especially in developing countries. In this regard, sewage treatment by coupling methanogenesis and autotrophic nitrogen removal, through a combined UASB-Anammox system, rises as an attractive option to reach this goal. However, and despite that these processes have been successfully applied for wastewater treatment, they have not been coupled for municipal wastewater treatment yet. More precisely, the implementation of Anammox for mainstream sewage treatment has not been studied extensively. Taking into account the advantages and benefits of both processes, their integration will raise valuable opportunities to develop an integrated anaerobic system for COD and DIN removal from sewage in urban zones. The combination of these processes in a treatment system offers important advantages such as (i) relatively high treatment efficiency for both carbon and nitrogen, (ii) low sludge emissions, (iii) relatively small footprint when compared to activated sludge systems, (iv) no need of external organic carbon source for nitrogen removal, and (v) a positive energy balance (Sliekers *et al.*, 2002; Siegrist *et al.*, 2008; van Loosdrecht, 2008; van Lier *et al.*, 2008).

At present, methanogenesis coupled to autotrophic nitrogen removal via an Anammox process is only applied for industrial wastewater (Abma *et al.*, 2010; Lackner *et al.*, 2014) and sludge reject water in sewage treatment. In this regard, the composition of industrial wastewater considerably differs from that of sewage, e.g. higher temperature, higher concentrations of nitrogen and minerals, and lower COD/N ratios. Hence, there are scientific and practical challenges that need to be addressed prior to assessing the feasibility of integrating and operating a combined UASB-Anammox system for mainstream municipal wastewater treatment. This research focuses on the study of these challenges that have not been investigated previously or need more investigation and that would have a significant impact in the successful integration of UASB and Anammox reactors. It is envisaged that this study will provide relevant information about the practical application of nitrogen removal from sewage, supporting the development of design and operational guidelines for sewage treatment via an integrated UASB-ANAMMOX system. Ultimately, the knowledge acquired will serve to develop a technology to reduce the nitrogen discharges from sewage into the environment.

1.4. Research objectives

This research has the following specific objectives. They have been defined based on the relevant challenges that have been identified:

1. To study the short-term effects on the Anammox process of organic carbon fractions that are generally present in UASB effluents, i.e. readily biodegradable COD (RBCOD) and slowly biodegradable COD (SBCOD), in dependence to the absolute nitrogen concentration and sewage temperature.

2. To assess the long-term effects of low sludge nitrogen loading rates and moderate to low temperatures on the Anammox process for DIN removal from a synthetic anaerobic medium. The envisaged results provide useful information for the operation of granular Anammox systems that are inoculated with Anammox granules, which are grown at high sludge nitrogen loading rate.

3. To evaluate the performance of the Anammox process in a closed sponge bed trickling filter (CSTF), as a low-cost post-treatment technology for DIN removal from UASB effluents.

4. To assess the feasibility of a one-stage Anammox process using natural air convection in sponge bed trickling filters (STF) as a low-cost post-treatment step for UASB effluents.

1.5. Outline of the thesis

The thesis is organized in six chapters. After this introductory **Chapter 1**, the following four chapters are devoted to present the research of each topic identified as a potential challenge to be solved or studied in order to achieve a successful integration of the UASB-Anammox system. The final chapter presents the outlook of the entire research. Figure 1.10 depicts the diverse themes studied in this research.

- **Chapter 2** presents an assessment of the short-term simultaneous effect of organic carbon sources, COD/N ratio and temperature on autotrophic nitrogen conversion by performing batch tests. Results are evaluated using the stoichiometric NO_2^--N/NH_4^+-N conversion ratios and DIN removal efficiencies. The envisaged results reflect the short-term response of anammox biomass on incoming organic matter that might be present in anaerobic effluents.

- In **Chapter 3** the long-term influence of a low nitrogen sludge loading rate (NSLR) on the Anammox process in a sequencing batch reactor (SBR) with granular Anammox biomass was evaluated. The research focused on the dynamics of both the metabolic properties and the morphology of the Anammox granular sludge that was used as inoculum. The Anammox granular sludge was grown in an Anammox reactor treating sludge reject water. The imposed SBR conditions simulate expected field conditions in which low NSLR are expected after anaerobic pre-treatment.

- **Chapter 4** describes the study of the Anammox process in a closed sponge bed trickling filter (CSTF) for anaerobic effluent post-treatment as a simple technology with low operational and maintenance requirements. Research on the immobilization and cultivation of Anammox bacteria in the CSTF reactors will provide useful information for developing an alternative DIN removal technology, which is characterized by a high biomass retention capacity. In addition, the study includes the effect of operational temperature on the

Anammox process in the CSTF reactors, in order to evaluate the applicable temperature range.

- **Chapter 5** describes the technical feasibility of single stage autotrophic nitrogen removal over nitrite using sponge-bed trickling filters (STF) under natural air convection. The research is focused on developing autotrophic NH_4^+-N removal in the STF reactors using activated sludge as inoculum, achieving a short start-up period. After a successful start-up, the STF reactors may be operated without mechanical aeration control systems, whereas a high nitrogen removal at a small footprint is expected.

- In **Chapter 6** provides an outlook with the technical, scientific and social repercussions of the findings of this study, as well as suggestions for additional research.

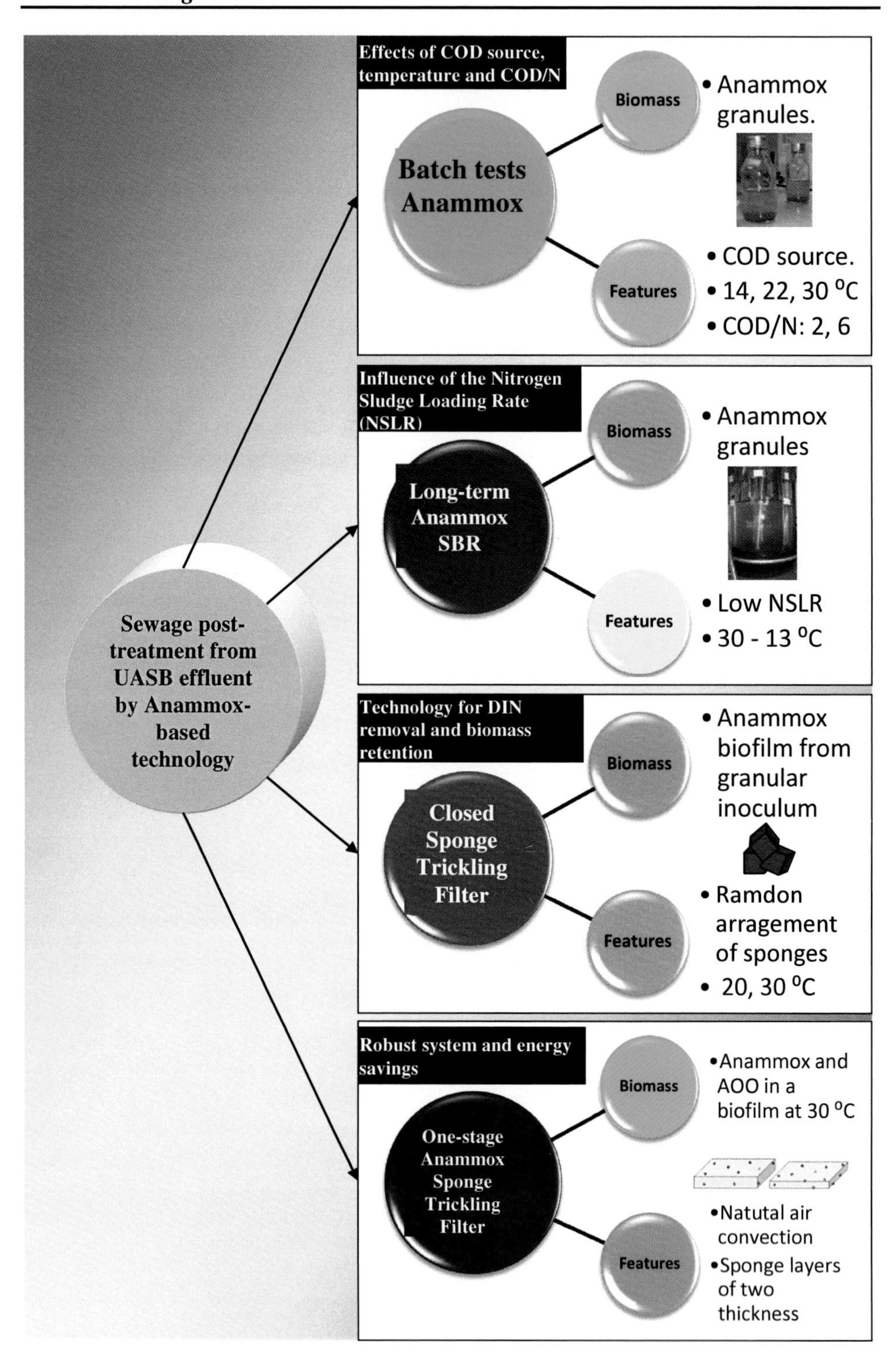

Figure 1.10. Schematic outline of the thesis.

References

Abma W.R., Driessen W., Haarhuis R., van Loosdrecht M.C.M., 2010. Upgrading of sewage treatment plant by sustainable and cost-effective separate treatment of industrial wastewater. Water Science and Technology, 61 (7), 1715-1722.

Ali M., Oshiki M., Awata T., Isobe K., Kimura Z., Yoshikawa H., Hira D., Kindalchi T., Satoh H., Fujii T., Okabe S., 2015. Physiological characterization of anaerobic ammonium oxidizing bacterium "Candidatus Jettenia caeni". Environmental Microbiology, 17 (6), 2172-2189.

Ali M. and Okabe S., 2015. Anammox-based technologies for nitrogen removal: Advances in process start-up and remaining issues. Chemosphere, 141, 144-153.

Almeida P.G.S., Chernicharo C.A.L., Souza C.L., 2009. Development of compact UASB/trickling filter systems for treating domestic wastewater in small communities in Brazil. Water Science and Technology, 59 (7), 1431-1439.

Almeida P.G.S., Marcus A.K., Rittmann B.E., Chernicharo C.A.L., 2013. Performance of plastic - and sponge - based trickling filters treating effluents from an UASB reactor. Water Science and Technology, 67 (5), 1034-1042.

Al-Omari A., Wett B., Nopens I., De Clipper H., Han M., Regmi P., Bott C., Murthy S., 2015. Model-based evaluation of mechanisms and benefits of mainstream shortcut nitrogen removal processes. Water Science and Technology, 71 (6), 840-847.

Awuah E. and Abrokwa K.A., 2008. Performance evaluation of the UASB sewage treatment plant at James Town (Mudor), Accra. Conference Proceedings, 33rd WED International Conference, Access to sanitation and safe water: Global partnerships and local actions. Accra, Ghana.

Bandara W.M.K.R.T.W., Satoh H., Sasakawa M., Nakahara Y., Takahashi M., Okabe S., 2011. Removal of residual dissolved methane gas in an upflow anaerobic sludge blanket reactor treating low-strength wastewater at low temperature with degassing membrane. Water Research, 45, 3533-3540.

Bartrolí A., Carrera J., Pérez J., 2011. Bioaugmentation as a tool for improving the start-up and stability of a pilot-scale partial nitrification biofilm airlift reactor. Bioresource Technology, 102, 4370-4375.

Bashkin V.N., Park S.U., Choi M.S., Lee C.B., 2002. Nitrogen budgets for the Republic of Korea and the Yellow Sea region. Biogeochemistry, 57/58, 387-403.

Billen G., Garnier J., 2007. River basin nutrient delivery to the coastal sea: Assessing its potential to sustain new production of non-siliceous algae. Mar. Chem., 106 (1-2 SPEC. ISS.), 148-160.

Björklund G., Burke J., Foster S., Rast W., Vallée D., van der Hoek W., Bernardini F., Cleveringa R., Cohen A., Faurès J.M., Koo-Oshima S., Kuonqui C., Mutandi R., Stracasto L., Le-Huu T., Winpenny J., 2009. Impacts of water use on water systems and the environment. In World Water Assessment Programme. The United Nations World Water Development Report 3: Water in a Changing World. UNESCO Publishing and London: Earthscan, pp. 127-149. http://www.unesco.org/water/wwap/wwdr/wwdr3/tableofcontents.shtml

Cao Y., Kwok B.H., Yong W.H., Chua S.C., Wah Y.L., Ghani Y.A.B.D., 2013. Mainstream partial nitritation-anammox nitrogen removal in the largest full-scale activated sludge process in Singapore: process analysis. Conference proceedings, WEF/IWA Nutrient Removal and Recovery 2013: Trends in Resource Recovery and Use. Vancouver, Canada.

Carvajal-Arroyo J.M., Sun W., Sierra-Alvarez R., Field J.A., 2013. Inhibition of anaerobic ammonium oxidizing (anammox) enrichment cultures by substrates, metabolites and common wastewater constituents. Chemosphere, 91, 22-27.

Chernicharo C.A.L. and Nascimento M.C.P., 2001. Feasibility of a pilot-scale UASB/trickling filter system for domestic sewage treatment. Water Science and Technology, 44 (4), 221-228.

Chernicharo C.A.L., 2006. Post-treatment options for the anaerobic treatment of domestic wastewater. Reviews in Environmental Science and Bio/Technology, 5, 73-92.

Chernicharo C.A.L., Almeida P.G.S., Lobato L.C.S., Couto T.C., Borges J.M., Lacerda Y.S., 2009. Experience with the design and start-up of two full-scale

UASB plants in Brazil: enhancements and drawbacks. Water Science and Technology, 60 (2), 507-515.

Chernicharo C.A.L., Almeida P.G.S., Lobato L.C.S., Rosa A.P., 2012. Anaerobic domestic wastewater treatment in Brazil: drawbacks, advances and perspectives. Water 21, October 2012, 24-26.

Chernicharo C.A.L., van Lier J.B., Noyola A., Bressani Ribeiro T., 2015. Anaerobic sewage treatment: state of the art, constraints and challenges. Reviews in Environmental Science and Bio/technology, 14 (4), 649-679.

Claros J., Serralta J., Seco A., Ferrer J., Aguado D., 2012. Real-time control strategy for nitrogen removal via nitrite in a SHARON reactor using pH and ORP sensors. Process Biochemistry, 47, 1510-1515.

Cydzik-Kwiatkowska A. and Wojnowska-Baryla I., 2015. Nitrogen-converting communities in aerobic granules at different hydraulic retention times (HRTs) and operational modes. World J. Microbiol. Biotechnol., 31, 75-83.

Dang H., Zhou H., Zhang Z., Yu Z., Hua E., Liu X., Jiao N., 2013. Molecular detection of *Candidatus* Scalindua pacifica and environmental responses of sediment anammox bacterial community in the Bohai Sea, China. PLoS ONE, 8:e61330. doi: 10.1371/journal.pone.0061330.

Dapena-Mora A., Fernández I., Campos J.L., Mosquera-Corral A., Méndez R., Jetten M.S.M., 2007. Evaluation of activity and inhibition effects on Anammox process by batch tests based on the nitrogen gas production. Enzyme and Microbial Technology, 40, 859-865.

de Graaff M.S., Zeeman G., Temmink H., van Loosdrecht M.C.M., Buisman C.J.N., 2011. Autotrophic nitrogen removal from black water: calcium addition as a requirement for settleability. Water Research, 45, 63-74.

de Kreuk M.K., 2006. Aerobic granular sludge: scaling up a new technology. PhD dissertation. Delft University of Technology, the Netherlands. ISBN: 978-90-9020767-4.

de Mes T., Hyde R., Hyde K., 2004. Anaerobic first for Ghana. Water 21, 30-31.

Dias D.F.C., Possmoser-Nascimento T.E., Rodrigues V.A.J., von Sperling M., 2014. Overall performance evaluation of shallow maturation ponds in series treating UASB reactor effluent: Ten years of intensive monitoring of a system in Brasil. Ecological Engineering, 71, 206-214.

Ekama G.A., Wentzel M.C., 2008a. Organic Material Removal. In Biological Wastewater Treatment: Principles, Modelling and Design (M. Henze, M.C.M. van Loosdrecht, G.A. Ekama and D. Brdjanovic, eds.), IWA Publishing, London, UK, pp. 76.

Ekama G.A., Wentzel M.C., 2008b. Nitrogen Removal. In Biological Wastewater Treatment: Principles, Modelling and Design (M. Henze, M.C.M. van Loosdrecht, G.A. Ekama and D. Brdjanovic, eds.), IWA Publishing, London, UK, pp. 111-112.

FAO, 2009. Consumption data for livestock and fish primary equivalents. FAOSTAT online statistical service. Rome: FAO. http://faostat.fao.org/

Ferreira da Costa J., Pinto Martins W.L., Seidl M., von Sperling M., 2015. Role of vegetation (*Typha latifolia*) on nutrient removal in a horizontal subsurface-flow constructed wetland treating UASB reactor-trickling filter effluent. Water Science and Technology, 71 (7), 1004-1010.

Foresti E., 2002. Anaerobic treatment of domestic sewage: established technologies and perspectives. Water Science and Technology, 45 (10), 181-186.

Fuchsman C.A., Staley J.T., Oakley B.B., Kirkpatrick J.B., Murray J.W., 2012. Free-living and aggregate-associated *Planctomycetes* in the Black Sea. FEMS Microbiol. Ecol., 80, 402-416.

Gao H., Scherson Y.D., Wells G.F., 2014. Toward energy neutral wastewater treatment: methodology and state of art. Environ. Sci. Processes Impacts, 16, 1223-1246.

Garnier J., Beusen A., Thieu V., Billen G., Bouwman L., 2010. N:P:Si nutrient export ratios and ecological consequences in coastal seas evaluated by the ICEP approach. Global Biogeochem. Cycles, 24 (2): GB0A05.

Ge S., Pen Y., Qiu S., Zhu A., Ren N., 2014. Complete nitrogen removal from municipal wastewater via partial nitrification by appropriately alternating anoxic/aerobic conditions in a continuous plug-flow step feed process. Water Research, 55, 95-105.

Giesen A., de Bruin L.M.M., Niermans R.P., van der Roest H.F., 2013. Advancements in the application of aerobic granular biomass technology for sustainable treatment of wastewater. Water Practice & Technology, 8 (1), 47-54.

Gijzen H.J., 2002. Anaerobic digestion for sustainable development: a natural approach. Water Science and Technology, 45 (10), 321-328.

Giraldo E., Pena M., Chernicharo C.A.L., Sandino J., Noyola A., 2006. Anaerobic sewage treatment technology in Latin-America: A selection of 20 years of experiences. Proceedings of the Water Environment Federation 12/2006. San Diego, CA. USA.

Gujer W., 2010. Nitrification and me - A subjective review. Water Research, 44, 1-19.

Harada H., 2014. UASB-DHS integrated system: a sustainable sewage treatment technology. In: Proc. of International Workshop on "UASB-DHS integrated system – a Sustainable Sewage Treatment Technology", 16-18 October, 2015, Agra, New Delhi & Agra, India.

Hatamoto M., Yamamoto H., Kindaichi T., Ozaki N., Ohashi A., 2010. Biological oxidation of dissolved methane in effluents from anaerobic reactors using a down-flow hanging sponge reactor. Water Research, 44, 1409-1418.

Heffernan B., van Lier J.B., van der Lubbe J., 2011. Performance review of large scale up-flow anaerobic sludge blanket sewage treatment plants. Water Science and Technology, 63 (1), 100-107.

Hellinga C., Schellen A.A.J.C., Mulder J.W., van Loosdrecht M.C.M., Heijnen J.J., 1998. The SHARON process: An innovative method for nitrogen removal from ammonium-rich waste water. Water Science and Technology, 37 (9), 135-142.

Hendrickx T.L.G., Kampman C., Zeeman G., Temmink H., Hu Z., Kartal B., Buisman C.J.N., 2014. High specific activity for anammox bacteria enriched from activated sludge at 10^0C. Bioresource Technology, 163, 214-221.

Henze M. and Comeau Y., 2008. Wastewater Characterization. In Biological Wastewater Treatment: Principles, Modelling and Design (M. Henze, M.C.M.

van Loosdrecht, G.A. Ekama and D. Brdjanovic, eds.), IWA Publishing, London, UK, pp. 35.

Henze M., van Loosdrecht M.C.M., Ekama G.A., Brdjanovic D., 2008. Wastewater Treatment Development. In Biological Wastewater Treatment: Principles, Modelling and Design (M. Henze, M.C.M. van Loosdrecht, G.A. Ekama and D. Brdjanovic, eds.), IWA Publishing, London, UK, pp. 1-2.

Hippen A., Rosenwinkel K.H., Baumgarten G., Seyfried C.F., 1997. Aerobic deammonification: a new experience in the treatment of wastewaters. Water Science and Technology, 35, 111-120.

Hong Y.G., Li M., Cao H., Gu J.D., 2011. Residence of habitat-specific anammox bacteria in the deep-sea subsurface sediments of the South China Sea: analyses of marker gene abundance with physical chemical parameters. Microb. Ecol., 62, 36-47.

Howarth R., Billen G., Swaney D., Townsend A., Jaworski N., 1996. Regional nitrogen budgets and riverine N and P fluxes for the drainages to the North Atlantic Ocean: Natural and human influences. Biogeochemistry, 35, 75-139.

Howarth R., Boyer E.W., Pabich W.J., Galloway J.N., 2002. Nitrogen use in the United States from 1961-2000 and potential future trends. Ambio., 31 (2), 88-96.

Howarth R., Ramakrishna K., Choi E., Elmgren R., Martinelli L., Mendoza A., Moomaw W., Palm C., Roy R., Scholes M., Zhao-Liang Z., Etchvers J., Tiessen H., 2005. Nutrient Management. In Millennium Ecosystem Assessment. Ecosystems and Human Well-being: Policy Responses (Chopra K., Leemans R., Kumar P., Simons H., eds.). Vol. 3, chapter 9, pp. 295-311. http://maweb.org/en/Global.aspx

Howarth R. and Marino R., 2006. Nitrogen as the limiting nutrient for eutrophication in coastal marine ecosystems: Evolving views over three decades. Limnol. Oceanogr., 51 (1, part 2), 364-376.

Hu Z., Lotti T., van Loosdrecht M., Kartal B., 2013. Nitrogen removal with the anaerobic ammonium oxidation process. Biotechnol. Lett., 35, 1145-1154.

Isanta E., Reino C., Carrera J., Pérez J., 2015. Stable partial nitritation for low-strength wastewater at low temperature in an aerobic granular reactor. Water Research, 80, 149-158.

Jetten M.S.M., Strous M., van de Pas-Schoonen K.T., Schalk J., van Dongen U.G.J.M., van De Graaf A.A., Logemann S., Muyzer G., van Loosdrecht M.C.M., Kuenen J.G., 1999. The anaerobic oxidation of ammonium. FEMS Microbiology Reviews, 22, 421-437.

Jin R.C., Yang G.F., Zhang Q.Q., Ma C., Yu J.J., Xing B.S., 2013. The effect of sulfide inhibition on the ANAMMOX process. Water Research, 47, 1459-1469.

Jørgensen D., 2010. Local government responses to urban river pollution in late medieval England. Water Hist. 2, 35-52.

Joss A., Salzgeber D., Eugster J., Koning R., Rottermann K., Burger S., Fabijan P., Leumann S., Mohn J., Siegrist H., 2009. Full-scale nitrogen removal from digester liquid with Partial Nitritation and ANAMMOX in one SBR. Environ. Sci. Technol., 43, 5301-5306.

Kartal B., Rattray J., van Niftrik L.A., van de Vossenberg J., Schmid M.C., Webb R.I., Schouten S., Fuerst J.A., Damsté J.S., Jetten M.S.M., Strous M., 2007. Candidatus "Anammoxoglobus propionicus" a propionate oxidizing species of anaerobic ammonium oxidizing bacteria. Systematic and Applied Microbiology, 30, 39-49.

Kartal B., van Niftrik L., Rattray J., van de Vossenberg J.L.C.M., Schmid M., Sinninghe Damsté J., Jetten M.S.M., Strous M., 2008. *Candidatus* 'Brocadia fulgida': an autoflorescent anaerobic ammonium oxidizing bacterium. FEMS Microbiol. Ecol., 63, 46-55.

Kartal B., Kuenen J.G. and van Loosdrecht M.C.M., 2010. Sewage Treatment with Anammox. Science 38, 702-703.

Kartal B., van Niftrik L., Keltjens J.T., Op den Camp H.J.M., Jetten M.S.M., 2012. Anammox-growth physiology, cell biology and metabolism. In: Advances in Microbial Physiology (Poole R.K., ed.), Elsevier Ltd., Vol. 60, pp. 211-262.

Khan A.A., Gaur R.Z., Mehrotra I., Diamantis V., Lew B., Kazmi A.A., 2014. Performance assessment of different STPs based on UASB followed by aerobic post treatment systems. J. Environ. Health Sci. Eng., 12 (43), 1-13.

Khanal S.M., 2008. Anaerobic Biotechnology for Bioenergy Production. Principles and Applications. Wiley – Blackwell. USA, pp xiii.

Khramenkov S.V., Kozlov M.N., Kevbrina M.V., Dorofeev A.G., Kazakova E.A., Grachev V.A., Kuznetsov B.B., Polyakov D.Y., Nikolaev Y.A., 2013. A novel bacterium carrying out anaerobic ammonium oxidation in a reactor for biological treatment of the filtrate of wastewater fermented sludge. Microbiology, 82 (5), 628-636.

Kuenen J.G., 2008. Anammox bacteria: from discovery to application. Nature Reviews Microbiology, 6 (4), 320-326.

Kujawa-Roeleveld K., 2011. Natural Treatment Methods & Post Treatment after Anaerobic Pre-Treatment. Course: Principles of Anaerobic Wastewater Treatment. LeAF Foundation. Wageningen University, the Netherlands. March 23-25.

Kuypers M.M.M., Sliekers A.O., Lavik G., Schmid M., Jørgensen B.B., Kuenen J.G., Sinninghe Damsté J.S., Strous M., Jetten M.S.M., 2003. Anaerobic ammonium oxidation by anammox bacteria in the Black Sea. Nature, 422, 608-611.

Lackner S., Gilbert E.M., Vlaeminck S.E., Joss A., Horn H., van Loosdrecht M.C.M., 2014. Full-scale partial nitritation/anammox experiences - An application survey. Water Research, 55, 292-303.

Lettinga G., van Velsen A.F.M., Hobma S.W., de Zeeuw W.J., Klapwijk A., 1980. Use of the Upflow Sludge Blanket (USB) reactor concept for biological wastewater treatment. Biotech. Bioeng., 22, 699-737.

Lettinga G. and Hulshoff Pol L.W., 1991. UASB-process design for various types of wastewaters. Water Science and Technology, 24 (8), 87-107.

Li H., Chen S., Mu B.Z., Gu J.D., 2010. Molecular detection of anaerobic ammonium-oxidizing (anammox) bacteria in high-temperature petroleum reservoirs. Microb. Ecol., 60, 771-783.

Liu S., Yang F., Gong Z., Meng F., Chen H., Xue Y., Furukawa K., 2008. Application of anaerobic ammonium-oxidizing consortium to achieve completely autotrophic ammonium and sulfate removal. Bioresource Technology, 99, 6817-6825.

Lotti T., van der Star W.R.L., Kleerebezem R., Lubello C., van Loosdrecht M.C.M., 2012. The effect of nitrite inhibition on the anammox process. Water Research, 46, 2559-2569.

Lotti T., Kleerebezem R., Lubello C., van Loosdrecht M.C.M., 2014a. Physiological and kinetic characterization of a suspended cell anammox culture. Water Research, 60, 1 -14.

Lotti T., Kleerebezem R., Hu Z., Kartal B., Jetten M.S.M., van Loosdrecht M.C.M., 2014b. Simultaneous partial nitritation and anammox at low temperature with granular sludge. Water Research, 66, 111 -121.

Lotti T., Kleerebezem R., Hu Z., Kartal B., de Kreuk M.K., van Erp Taalman Kip C., Kruit J., Hendrickx T.L.G., van Loosdrecht M.C.M., 2015. Pilot-scale evaluation of anammox-based mainstream nitrogen removal from municipal wastewater. Environmental Technology, 36 (9), 1167-1177.

Mahmood Q., Zheng P., Cai J., Wu D., Hu B., Li J., 2007. Anoxic sulfide biooxidation using nitrite as electron acceptor. Journal of Hazardous Materials, 147, 249-256.

Mahmoud N., Amarneh M.N., Al-Sa'ed R., Zeeman G., Gijzen H., Lettinga G., 2003. Sewage characterization as a tool for the application of anaerobic treatment in Palestine. Environmental Pollution, 126, 115-122.

Malamis S., Katsou E., Frison N., Di Fabio S., Noutsopoulos C., Fatone F., 2013. Start-up of the completely autotrophic nitrogen removal process using low activity anammox inoculum to treat low strength UASB effluent. Bioresource Technology, 148, 467-473.

Malovanyy A., Yang J., Trela J., Plaza E., 2015. Combination of upflow anaerobic sludge blanket (UASB) reactor and partial nitritation/anammox moving bed biofilm reactor (MBBR) for municipal wastewater treatment. Bioresource Technology, 180, 144-153.

Mayorga E., Seitzinger S.P., Harrison J.A., Dumont E., Beusen A.H.W., Bouwman A.F., Fekete B.M., Kroeze C., van Drecht G., 2010. Global Nutrient Export from WaterSheds 2 (NEWS 2): Model development and implementation. Environmental Modelling & Software, 25, 837-853.

McCarty P.L, 1964. Anaerobic Waste Treatment Fundamentals. Public Works. September: 107-112, October: 123-126, November: 91-94, December: 95-99.

McCarty P.L, 2001. The development of anaerobic treatment and its future. Water Science and Technology, 44 (8), 157-176.

Millennium Ecosystem Assessment Board, 2005. Living Beyond Our Means. Natural Assets and Human Well-Being. Statement from the Board (Sarukhán J. and Whyte A. eds.), pp. 11. http://maweb.org/en/index.aspx

Monroy O., Famá G., Meraz M., Montoya L., Macarie H., 2000. Anaerobic digestion for wastewater treatment in Mexico: State of the technology. Water Research, 34 (6), 1803-1816.

Morales N., Val del Río A., Vázquez-Padín J.R., Méndez R., Mosquera-Corral A., Campos J.L., 2015. Integration of the Anammox process to the rejection water and main stream lines of WWTPs. Chemosphere, 140, 99-105.

Mulder A., Van De Graaf A.A., Robertson L.A., Kuenen J.G., 1995. Anaerobic ammonium oxidation discovered in a denitrifying fluidized bed reactor. FEMS Microb. Ecology, 16 (3), 177-184.

Mulder A., 2003. The quest for sustainable nitrogen removal technologies. Water Science and Technology, 48 (1), 67-75.

Nikolaev Y.A., Kozlov M.N., Kevbrina M.V., Dorofeev A.G., Pimenov N.V., Kallistova A.Y., Grachev V.A., Kazakova E.A., Zharkov A.V., Kuznetsov B.B., Patutina E.O., Bumazhkin B.K., 2015. *Candidatus* "Jettenia moscovienalis" sp. nov., a new species of bacteria carrying out anaerobic ammonium oxidation. Microbiology, 84 (2), 256-262.

NRC, 2000. Clean Coastal Waters: Understanding and Reducing the Effects of Nutrient Pollution, National Academy Press, Washington, D.C.

Noyola A., Padilla-Rivera A., Morgan-Sagastume J.M., Güereca L.P., Hernández-Padilla F., 2012. Typology of municipal wastewater treatment technologies in Latin America. Clean - Soil, Air, Water 40(9), 926–932.

Oenema O., 2011. Nitrogen in current European policies. In The European Nitrogen Assessment (M. A. Sutton, C.M. Howard, J.W. Erisman, G. Billen, A. Bleeker, P. Grennfelt, H. van Grinsven and B. Grizzetti, eds.), Cambridge University Press, UK, pp. 68.

Oleszkiewicz J.A., Barnard J.L., 2006. Nutrient Removal Technology in North America and the European Union: A Review. Water Qual. Res. J. Canada, 41 (4), 449-462.

Oshiki M., Shimokawa M., Fujii N., Satoh H., Okabe S., 2011. Physiological characteristics of the anaerobic ammonium-oxidizing bacterium "*Candidatus* Brocadia sinica". Microbiology, 157, 1706-1713.

Pérez J., Lotti T., Kleerebezem R., Picioreanu C., van Loosdrecht M.C.M., 2014. Outcompeting nitrite-oxidizing bacteria in single-stage nitrogen removal in sewage treatment plants: A model-based study. Water Research, 66, 208-218.

Pool, R., 2015. Coming clean: the future of sewage. Engineering & Technology. January, pp. 80-83. www.EandTmagazine.com

Procópio Pontes P. and Chernicharo C.A.L., 2011. Characterization and removal of specific organic constituents in an UASB-trickling-filter system treating domestic wastewater. Environmental Technology, 32 (3), 281-287.

Quan Z.X., Rhee S.K., Zuo J.E., Yang Y., Bae J.W., Park J.R., Lee S.T., Park Y.H., 2008. Diversity of ammonium-oxidizing bacteria in a granular sludge anaerobic ammonium-oxidizing (anammox) reactor. Environmental Microbiology, 10 (11), 3130-3139.

Rabalais N.N., Díaz R.J., Levin L.A., Turner R.E., Gilbert D., Zhang J., 2010. Dynamics and distribution of natural and human-caused hypoxia. Biogeosciences, 7, 585-619.

Rosa A.P., Lobato L.C.S., Chernicharo C.A.L., Martins D.C.R.B., Maciel F.M., Borges J.M., 2012. Improving performance and operational control of UASB reactors via proper sludge and scum discharge routines. Water Practice & Technology, 7 (3), doi: 10.2166/wpt.2012.046.

Rothrock M.J., Vanotti M.B., Szogi A.A., Gonzalez M.C. Fujii T., 2011. Long-term preservation of anammox bacteria. Appl. Microbiol. Biotechnol., 92, 147-157.

Russ L., Speth D.R., Jetten M.S.M., Op den Camp H.J.M., Kartal B., 2014. Interactions between anaerobic ammonium and sulfur-oxidizing bacteria in a laboratory scale model system. Environmental Microbiology, 16 (11), 3487-3498.

Sánchez de León E., 2001. Case Study: Sololá Guatemala Project on Treatment and Integral Use of Wastewater in Latin America: Actual situation and potential. Agreement IDRC-OPS/HEP/CEPIS 2000-2002 (In Spanish).

Sanz I. and Fdz-Polanco F., 1990. Low temperature treatment of municipal sewage in anaerobic fluidized bed reactors. Water Research, 24 (4), 463-469.

Schaubroeck T., De Clippeleir H., Weissenbacher N., Dewulf J., Boeckx P., Vlaeminck S.E., Wett B., 2015. Environmental sustainability of an energy self-sufficient sewage treatment plant: Improvements through DEMON and co-digestion. Water Research, 74, 166-179.

Schellinkhout A. and Osorio E., 1994. Long-term experience with the UASB technology for sewage treatment on large scale. Paper reprints of Seventh Inte. Symp. On Anaerobic Digestion, Cape Town, 23-27.

Schmid M., Twachtmann U., Klein M., Strous M., Juretschko S., Jetten M., Metzger J.W., Schleifer K.H, Wagner M., 2000. Molecular evidence for genus level diversity of bacteria capable of catalyzing anaerobic ammonium oxidation. Syst. Appl. Microbiol., 23, 93-106.

Schmid M., Walsh K., Webb R., Rijpstra W.I.C., van de Pas-Schoonen K., Verbruggen M.J., Hill T., Moffett B., Fuerst J., Schouten S., Sinninghe Damsté J.S., Harris J., Shaw P., Jetten M., Strous M., 2003. *Candidatus* "Scalindua brodae", sp. nov., *Candidatus* "Scalindua wagneri", sp. nov., two new species of anaerobic ammonium oxidizing bacteria. System Appl. Microbiol., 26, 529-538.

Seitzinger S.P., Harrison J.A., 2008. Land-Based Nitrogen Sources and Their Delivery to coastal Systems. In Nitrogen in the Marine Environment, 2nd Edition (Capone D., Bronk D.A., Mullholland M.R., Carpenter E., eds.). Academic Press, New York, pp. 469-510.

Seitzinger S.P., Mayorga E., Bouwman A.F., Kroeze C., Beusen A.H.W., Billen G., van Drecht G., Dumont E., Fekete B.M., 2010. Global river nutrient export: A scenario analysis of past and future trends. Global Biogeochemical Cycles, 24 (2):GB0A08.

Selman M., Greenhalgh S., Diaz R., Sugg Z., 2008. Eutrophication and Hypoxia in Coastal Areas: A Global Assessment of the State of Knowledge. Water Quality: Eutrophication and Hypoxia No. 1. Policy Note. World Resources Institute. Washington, D.C., pp. 2.
http://www.wri.org/publication/eutrophication-and-hypoxia-in-coastal-areas

Selman M., Greenhalgh S., 2009. Eutrophication: Sources and Drivers of Nutrient Pollution. Water Quality: Eutrophication and Hypoxia No. 2. Policy Note. World Resources Institute. Washington, D.C., pp. 2.
http://www.wri.org/publication/eutrophication-sources-and-drivers

Siegrist H., Salzgeber D., Eugster J., Joss A., 2008. Anammox brings WWTP closer to energy autarky due to increased biogas production and reduced aeration energy for N-removal. Water Science and Technology, 57 (3), 383-388.

Sliekers A.O., Derwort N., Campos Gomez J.L., Strous M., Kuenen J.G., Jetten M.S.M., 2002. Completely autotrophic nitrogen removal over nitrite in one single reactor. Water. Research, 36, 2475-2482.

Strous M., van Gerven E., Zheng P., Kuenen J.G., Jetten M.S.M., 1997. Ammonium removal from concentrated waste streams with the anaerobic ammonium oxidation (ANAMMOX) process in different reactor configurations. Water. Research, 31 (8), 1955-1962.

Strous M., Heijnen J.J., Kuenen J.G., Jetten M.S.M., 1998. The sequencing batch reactor as a powerful tool for the study of slowly growing anaerobic ammonium- oxidizing microorganisms. Appl. Microbiol., 50 (5), 589-596.

Strous M., Fuerst J.A., Kramer E.H., Logemann S., Muyzer G., van de Pas-Schoonen K.T., Webb R., Kuenen J.G., Jetten M.S.M., 1999. Missing lithotroph identified as new planctomycete. Nature, 400, 446-449.

Tchobanoglous G., Burton F.L. and Stensel H.D., 2003. Wastewater Engineering Treatment and Reuse. Fourth Edition. Metcalf & Eddy, Inc. Mc Graw Hill. New York, chapter 8, pp. 659-886.

Third K.A., Sliekers A.O., Kuenen J.G., Jetten M.S.M., 2001. The CANON System (Completely Autotrophic Nitrogen-removal Over Nitrite) under Ammonium Limitation: Interaction and Competition between Three Groups of Bacteria. System. Appl. Microbiol., 24 (4), 558-596.

Tokutomi T., Shibayama C., Soda S., Ike M., 2010. A novel control method for nitritation: The domination of ammonia-oxidizing bacteria by high concentrations of inorganic carbon in an airlift-fluidized bed reactor. Water Research, 44, 4195-4203.

Trela J., Plaza E., Szatkowska B., Hultman B., Bosander J., Dahlberg A.G., 2004. Deammonifakation som en ny process för behandling av avloppsströmmar med hög kvävehalt (Deammonification as a new process for treatment of wastewater with a high nitrogen content). Vatten, 60 (2), 119-127.

United Nations, 2015. The Millennium Development Goals Report 2015, Summary. Published by the United Nations Department of Economic and Social Affairs (DESA), pp. 8.
http://www.un.org/millenniumgoals/2015_MDG_Report/pdf/MDG%202015%20Summary%20web_english.pdf

United Nations, 2014. World Urbanization Prospects: The 2014 Revision, Highlights (ST/ESA/SER.A/352). Published by the United Nations Department of Economic and Social Affairs (DESA), Population Division, pp. 1.
http://esa.un.org/unpd/wup/highlights/wup2014-highlights.pdf

UNEP-GEMS/Water Programme, 2008. Water Quality for Ecosystem and Human Health 2nd Edition. UN/GEMS Water Programme Office, Canada, pp. 16-17.
http://www.gemswater.org/publications/index-e.html

van de Graaf A.A., Mulder A., de Bruijn P., Jetten M.S.M., Robertson L.A., Kuenen J.G., 1995. Anaerobic Oxidation of Ammonium Is a Biologically Mediated Process. Applied and Environmental Microbiology, 61 (4), 1246-1251.

van de Vossenberg J., Woebken D., Maalcke W.J., Wessels H.J., Dutilh B.E, Kartal B., Janssen-Megens E.M., Roeselers G., Yan J., Speth D., Gloerich J., Geerts W., van der Biezen E., Pluk W., Francoijs K-J., Russ L., Lam P., Malfatti S.A., Tringe S.G., Haaijer S.C.M., Op den Camp H.J.M., Stunnenberg H.G., Amann R., Kuypers M.M.M., Jetten M.S.M., 2013. The metagenome of the marine anammox bacterium 'Candidatus Scalindua profunda' illustrates the versatility of this globally important nitrogen cycle bacterium. Environ. Microbiol., 15 (5), 1275-1289.

van der Star W.R.L., Abma W.R., Blommers D., Mulder J.W., Tokutomi T., Strous M., Picioreanu C., van Loosdrecht M.C.M., 2007. Startup of reactors for anoxic ammonium oxidation: Experiences from the first full-scale anammox reactor in Rotterdam. Water Research, 41, 4149-4163.

van der Star W.R.L., Miclea A.I., van Dongen U.J.G.M., Muyzer G., Picioreanu C., van Loosdrecht M.C.M., 2008. The membrane bioreactor: a novel tool to grow anammox bacteria as free cells. Biotechnol. Bioeng., 41, 4149-4163.

van Dongen U., Jetten M.S.M., van Loosdrecht M.C.M., 2001. The SHARON-Anammox process for treatment of ammonium rich wastewater. Water Science and Technology, 44 (1), 153-160.

van Haandel A.C. and Lettinga G., 1994. Anaerobic Sewage Treatment. A practical Guide for Regions with a Hot Climate. John Wiley and Sons Ltd, Chichester.

van Hulle S.W.H., Vandeweyer H.J.P., Meesschaert B.D., Vanrolleghem P.A., Dejans P., Dumoulin A., 2010. Engineering aspects and practical application of autotrophic nitrogen removal from nitrogen rich streams. Chemical Engineering Journal, 162, 1-20.

van Lier J.B., Tilche A., Ahring B.K., Macarie H., Moletta R., Dohanyos M., Hulshoff Pol L.W., Lens P., Verstraete W., 2001. New perspectives in anaerobic digestion. Water Science and Technology, 43 (1), 1-18.

van Lier J.B., 2007. Current and future trends in anaerobic digestion: diversifying from waste (water) treatment to resource oriented conversion techniques. In:

Proceedings of the 11th IWA-International Conference on Anaerobic Digestion, Brisbane, September 23-27.

van Lier J.B., Mahmoud N., Zeeman G., 2008. Anaerobic Wastewater Treatment. In Biological Wastewater Treatment: Principles, Modelling and Design (M. Henze, M.C.M. van Loosdrecht, G.A. Ekama and D. Brdjanovic, eds.), IWA Publishing, London, UK, pp. 415-456.

van Lier J.B., Vashi A., van der Lubbe J., Heffernan B., 2010. Anaerobic sewage treatment using UASB reactors: engineering and operational aspects. In Environmental Anaerobic technology; Applications and New Developments (H.H.P. Fang ed.), World Scientific, Imperial College Press, London, UK, pp. 59-89.

van Lier J.B., van der Zee F. P., Frijters C.T.M.J., Ersahin M. E., 2015. Celebrating 40 Years Anaerobic Sludge Bed Reactors for Industrial Wastewater Treatment. Reviews in Environmental Science and Bio/technology, 14(4), 681-702.

van Loosdrecht M.C.M., 2008. Innovative Nitrogen Removal. In Biological Wastewater Treatment: Principles, Modelling and Design (M. Henze, M.C.M. van Loosdrecht, G.A. Ekama and D. Brdjanovic, eds.), IWA Publishing, London, UK, pp. 139-154.

Vieira P.C., von Sperling M., Nogueira L.C.M., Assis B.F.S., 2013. Performance of a novel trickling filter for the post-treatment of anaerobic effluents from small communities. Water Science and Technology, 67(12), 2746-2752.

von Sperling M. and Chernicharo C.A.L., 2005. Biological Waste-water Treatment in Warm Climate Regions. IWA Publishing, London 1452.

Wett B., 2007. Development and implementation of a robust deammonification process. Water Science and Technology, 56 (7), 81-88.

Wett B., Omari A., Podmirseg S.M. Han M., Akintayo O., Gómez Brandón M., Murthy S., Bott C., Hell M., Takacs I., Nyhuis G., O'Shaughnessy M., 2013. Going for mainstream deammonification from bench to full scale for maximized resource efficiency. Water Science and Technology, 68 (2), 283-289.

Winkler M.K.H., Kleerebezem R., van Loosdrecht M.C.M., 2012. Integration of anammox into the aerobic granular sludge process for main stream wastewater treatment at ambient temperatures. Water Research, 46, 136-144.

Woebken D., Lam P., Kuypers M.M.M., Naqvi S.W.A., Kartal B., Strous M., Jetten M.S.M., Fuchs B.M., Amann R., 2008. A microdiversity study of anammox bacteria reveals a novel *Candidatus* Scalindua phylotype in marine oxygen minimum zones. Environmental Microbiology, 10 (11), 3106-3119.

WRC, 2007. Integrated Water Resource Management Plan Guidelines for Local Authorities. Water Research Commission. Department of Water Affairs and Forestry. Republic of South Africa. Report N° TT 304/07, pp. 11. http://www.pacificwater.org/userfiles/file/IWRM%20Planning%20Guidelines %20for%20Local%20Authorities.pdf

Wyffles S., Boeckx P., Pynaert K., Verstraete W., Van Cleemput O., 2003. Sustained nitrite accumulation in a membrane-assisted bioreactor (MBR) for the treatment of ammonium-rich wastewater. J. Chem. Technol. Biotechnol., 78 (4), 412-419.

Xing G.X., Zhu Z.I., 2002. Regional nitrogen budgets for China and its major watersheds. Biogeochemistry, 57/58, 405-427.

Yu H., Tay J-H., Wilson F., 1997. A sustainable municipal wastewater treatment process for tropical and subtropical regions in developing countries. Water Science and and Technology, 35 (9), 191-198.

Short-term effects of organic carbon source, chemical oxygen demand/N ratio and temperature on autotrophic nitrogen removal

Contents

This chapter is based on:

Sánchez Guillén J.A., Yimman Y., Lopez Vazquez C.M., Brdjanovic D., van Lier J.B., 2014. Effects of organic carbon source, chemical oxygen demand/N ratio and temperature on autotrophic nitrogen removal. Water Science and Technology, 69 (10), 2079-2084.

Abstract

To assess the feasibility of the Anammox process as a cost-effective post-treatment step for anaerobic sewage treatment, the simultaneous effects of organic carbon source, COD/N ratio, and temperature on autotrophic nitrogen removal was studied. In batch experiments, three operating conditions were evaluated at 14, 22 and 30°C, and at COD/N ratios of 2 and 6. For each operating condition, containing 32 ± 2 mg-NH_4^+-N/L and 25 ± 2 mg-NO_2^--N, three different substrate combinations were tested to simulate the presence of readily biodegradable and slowly biodegradable organic matter (RBCOD and SBCOD, respectively): (i) acetate (RBCOD), (ii) starch (SBCOD); and, (iii) acetate + starch. The observed stoichiometric NO_2^--N/NH_4^+-N conversion ratios were in the range of 1.19-1.43 and the single or simultaneous presence of acetate and starch did not affect the Anammox metabolism. High autotrophic nitrogen removal was observed at 22°C (77-84%) and 30°C (73-79%), whereas there was no nitrogen removal at 14°C; the Anammox activity was strongly influenced by temperature, irrespective the COD source and COD/N ratios applied.

2.1. Introduction

Sewage is the main cause of point-source nitrogen (N) pollution. The Dissolved Inorganic Nitrogen (DIN) is the most abundant N-compound, representing up to 75% of the total nitrogen content (Henze and Comeau, 2008). The negative effects caused by the discharge of N-compounds on the environment mainly include eutrophication and hypoxia in water bodies. If the current removal efficiencies of nitrogen from sewage remain, the discharge of DIN will continue to increase sharply until 2030 (Seitzinger and Harrison, 2008). This scenario has increased the need to have more efficient DIN removal processes in sewage treatment plants and to develop and implement cost-effective N-removal technologies, such as the Anammox process that requires less aeration, no organic carbon source and generates fewer greenhouse gas emissions.

So far, the Anammox process has been applied successfully for N-removal in side-stream treatment lines at sewage treatment plants, e.g. in reject water from sludge handling facilities. But, its potential application as a post-treatment technology for N removal from effluents of an anaerobic sewage treatment process, e.g. effluents from upflow anaerobic sludge blanket (UASB) reactors, depends on the environmental and operational conditions, and sewage characteristics (Chernicharo *et al.*, 2015). These conditions and characteristics commonly differ from those found in side-stream lines, i.e. the presence of organic carbon, lower N-concentrations, and lower temperatures.

Certain Anammox bacteria are able to carry out the oxidation of the organic acid formate, acetate and propionate coupled with the reduction of nitrate or nitrite. This nitrate reduction via Anammox bacteria produces nitrite and ammonium. From these compounds, Anammox microorganisms generate dinitrogen gas (Kartal *et al.*, 2007, 2008). It has been proposed that the Anammox bacteria enriched in reactors fed with propionate and acetate are able to out-compete heterotrophic denitrifiers based on the mechanism of the substrate affinities (Kartal *et al.*, 2012). On the other hand, Veys *et al.* (2010) have applied an extension of the Activated Sludge Model (ASM) for the study of the Anammox process. The simulation study was performed at temperatures between 15 and 40°C. From the model-based analysis, they concluded that, provided there are optimum conditions for Anammox growth, a simultaneous removal of COD and N can be achieved by the cooperation between the ammonium oxidizing bacteria, Anammox and heterotrophic bacteria, particularly at low COD loads.

Taking into consideration the biodegradable COD fractions and the DIN concentration in sewage of medium strength (Henze and Comeau, 2008), as well as the removal efficiencies of these compounds using large scale UASB reactors (Heffernan *et al.*, 2011), it is expected that the effluent of the UASB reactors could contain approximately 60 mg of DIN /L and COD/N ratios between 2 and 6 (80% and 50% removal efficiency of biodegradable COD, respectively). As such, further research is necessary to elucidate the effects of key wastewater characteristics present in the UASB effluents, such as the organic carbon fractions in terms of their biodegradability (readily biodegradable COD: RBCOD, slowly biodegradable COD: SBCOD) and their relationship with the nitrogen concentrations and sewage temperature. For instance, Procópio Pontes and Chernicharo (2011) have used a UASB-trickling filter system for the treatment of sewage. They have reported carbohydrates (SBCOD) in the effluent of the UASB in the range of 19.7 (± 1.9) mg/L and 12.2 (± 4.8) mg/L.

The temperature influence on Anammox activity has been investigated by several authors. The optimum temperature for Anammox growth has been assessed between 30 and 40°C (Strous *et al.*, 1999). Dosta *et al.* (2008) performed short and long-term experiments to assess the influence of temperature on Anammox activity. Short-term experiments also showed a maximum activity at 35-40 °C. The long-term experiments were performed at temperatures between 30 to 15°C and the system was successfully operated down to 18°C; the stability of the system was lost when the temperature was decreased to 15°C.

In order to analyze the effect of temperature on nitrogen removal by Anammox, it is necessary to consider physiological key aspects about the growth temperature of the Anammox strain used in the research. In this sense, the Anammox group *Brocadia* is one of the six known genera of Anammox bacteria (Jetten *et al.*, 2010; Khramenkov *et al.*, 2013). So far, four species comprise the genus *Brocadia*: *Candidatus Brocadia caroliniensis, Candidatus Brocadia fulgida, Candidatus Brocadia anammoxidans* and *Candidatus Brocadia sinica*. The growth temperature of the last two species has been reported to range between 20 to 43°C (Strous *et al.*, 1999) and 25 to 45°C (Oshiki *et al.*, 2011), respectively. For the *Candidatus Brocadia fulgida*, the growth temperature has been tested within different ranges ranging from as low as 16 up to 35°C (Kartal *et al.*, 2008; Park *et al.*, 2010; Rattray *et al.*, 2010; Figueroa *et al.*, 2012; Winkler *et al.*, 2012; Liu *et al.*, 2013 and Puyol *et al.*, 2013).

It is vital to study the combined effects of the features previously described to understand the interaction and competition among microbial communities, e.g. denitrifying ordinary heterotrophs and Anammox bacteria, and to promote the ones that favor Anammox metabolism. Thus, the objective of this research is to perform preliminary tests to assess the simultaneous short-term effects of organic carbon sources, COD/N ratio and temperature on autotrophic nitrogen conversion by performing batch tests, based on the stoichiometric NO_2^--N/NH_4^+-N conversion ratios and nitrogen removal efficiencies. The results will contribute toward assessing, as a first approach, the feasibility for the potential implementation of the Anammox process as a post-treatment-stream step of the UASB reactors treating sewage, as a function of the effluent characteristics expected in different climates.

2.2. Materials and methods

2.2.1. Inoculum and substrates

The Anammox sludge used as inoculum to perform the batch activity tests was obtained from the Dokhaven-Sluisjesdijk wastewater treatment plant (Rotterdam, the Netherlands). This inoculum has the same origin and characteristics as the biomass used by Lotti *et al.*, (2012) with a maximum specific Anammox activity (MSAA) of 0.458 g-N_2-N/g-VSS d and a granule diameter of 1.1 ± 0.2 mm (94% of the granules). Based on molecular techniques, the dominant Anammox microorganisms were *Candidatus Brocadia fulgida* (Lotti, 2013). The synthetic autotrophic substrate applied in the experiments was modified from the substrate used by van de Graaf *et al.*, (1997) as presented in Table 2.1. The COD source for acetate was supplied as $CH_3COONa{\cdot}3H_2O$ and for starch as $[C_6H_{10}O_5]$; their concentrations were adjusted according to the COD/N ratio required, i.e. either 2 or 6.

2.2.2. Anammox batch tests

The nitrogen removal efficiencies and the NO_2^--N/NH_4^+-N conversion ratios were estimated under the operating conditions of interest through the execution of short-term batch tests in 310 mL vials. The Anammox sludge, the synthetic autotrophic substrate, the COD solutions and a washing buffer solution (0.14 g/L of KH_2PO_4 and 0.75 g/L of K_2HPO_4) were left overnight at the corresponding temperature being studied. Prior to the execution of the tests, the Anammox sludge was washed three times with the buffer solution with the aim of removing any residual organic carbon or nitrogen compounds. Afterwards, the supernatant was discarded. The synthetic autotrophic substrate containing 32 ± 2 mg-NH_4^+-N /L and 25 ± 2 mg-NO_2^--N/L was used as the nitrogen source and different concentrations of acetate and starch were added to simulate the presence of RBCOD and SBCOD. Tests were carried out in duplicate under 4 different combinations as follows: (i) control: only N-compounds without any organic substrate, (ii) acetate: N-compounds plus acetate (as RBCOD source), (iii) acetate + starch: N-compounds plus acetate and starch (25% and 75% on a COD basis as RBCOD and SBCOD sources, respectively) and (iv) starch: N-compounds plus starch (as SBCOD). Three different experimental conditions were tested at 14, 22 and 30°C, at COD/N ratios of 2 and 6. Before the experiments, the pH level in the vials was adjusted to 7.8-8.01 using 0.1M HCl and/or 0.1 M NaOH. Then, samples were collected to analyze the initial concentration of NH_4^+-N, NO_2^--N, NO_3^--N, COD and the mixed liquor volatile suspended solids (MLVSS). After the addition of the Anammox biomass and substrates, the average MLVSS concentration in the vials was approximately 1.5 g/L with a final working volume of 200 mL. The vials were closed with a gas-tight coated septum to avoid oxygen intrusion and the headspace was sparged with Helium gas for 5 minutes at a flow pressure of 0.2 bars to remove any remaining oxygen from the gas phase. Later on, the vials were incubated at the selected temperature and continuously stirred for 6 hours. At the end of the experiments, the final pH was determined and samples were taken to determine the final concentrations of NH_4^+-N, NO_2^--N and NO_3^--N.

Table 2.1. Synthetic autotrophic substrate

Autotrophic substrate		Trace solution	
Compound	Concentration (mg/L)	Compound	Concentration (mg/L)
NH_4^+-N	30	Mg EDTA	1500
NO_2^--N	30	$ZnSO_4 \cdot 7H_2O$	430
		$CoCl_2 \cdot 6H_2O$	240
Na EDTA	11.43	$MnCl_2 \cdot 4H_2O$	990
		$CuSO_4 \cdot 5H_2O$	250
$FeSO_4 \cdot 7H_2O$	11.43	$Na_2MoO_4 \cdot 2H_2O$	220
		$NiCl_2 \cdot 6H_2O$	190
$NaHCO_3$	11.88	$Na_2SeO_4 \cdot 10H_2O$	210
		H_3BO_3	14
Trace solution	1.25 mL/L	$NaWO_4 \cdot 2H_2O$	50

2.2.3. Analytical Methods

Prior to analyses, samples collected from the vials were centrifuged at 3600 rpm for a period of 20 minutes to separate the solid particles. For COD determination, 5 mL of the supernatant were taken and the rest of the volume was filtered using 0.45 μm Ion Chromatography (IC) Acrodisc® Syringe Filters (Pall Corporation, the Netherlands). Filtered samples were used for the determination of NH_4^+-N, NO_2^--N and NO_3^--N. The concentration of NH_4^+-N was determined spectrophotometrically according to NEN 6472 (NEN, 1983). MLVSS, NO_2^--N (spectrophotometric method) and COD (closed reflux colorimetric method) were analyzed according to the Standard Methods for the Examination of Water & Wastewater (2012). The DIONEX ICS-1000 ion chromatography system (Thermo Scientific, USA) was utilized for the analysis of NO_3^--N and pH was measured with a Metrohm 691 pH-meter (Metrohm, Switzerland).

2.3. Results and discussion

2.3.1. Nitrogen removal

Considering the overall objective of our study to develop an Anammox technology for the main-stream of a sewage treatment system under (sub) tropical conditions, in current experiments the effect of temperature on Anammox activity was studied at 30°C, 22°C and 14°C. The results of the batch tests show that Anammox bacteria were strongly influenced by temperature regardless of COD source and COD/N ratios applied. High values of total nitrogen removal (Table 2.2) were observed at 30°C (73-79%) and 22°C (77-84%), but there was no nitrogen removal at 14°C.

Overall, and despite Anammox bacteria not being active at 14°C, it was interesting to note the lack of the denitrification activity over nitrite by the ordinary heterotrophic organisms (OHO) at 14°C. It could be argued that at 14°C, heterotrophic denitrification would occur in the control using the COD available in the sludge matrix after washing the sludge, or based on the acetate and/or starch added as the source of organic matter in the other samples. However, at 14°C the denitrification kinetics rate (K_1) will be reduced to 5% of the activity reported at 30°C (Ekama and Wentzel, 2008), which could explain the lack of denitrifying activity. On the other hand, and possibly, denitrifying organisms were simply not present in the inoculum.

Total nitrogen removal efficiencies in the presence of the different COD sources at 30 and 22°C were similar to the average control values at both COD/N ratios (2 and 6 gCOD/gN), suggesting that the Anammox bacteria were not affected by the presence of COD sources during the batch experiments. These results imply that temperatures in the range of 22-30°C at the short-term do not affect the activity of Anammox bacteria or, at least, do not favor the activity of other microorganisms (e.g. ordinary heterotroph organisms) in detriment to Anammox, regardless of whether a carbon source is present. Long-term studies are needed to validate these observations and to assess the effects on the microbial populations and their interactions. With regard to the effect of the organic carbon source, the NO_2^--N was fully removed at 30°C and 22°C, with or without an organic source added. Overall, and although other factors may also influence the carbon source effects on the performance of the Anammox process, the observations suggest that the presence of starch (SBCOD) and acetate (RBCOD) do not appear to affect the metabolism of Anammox consortia.

2.3.2. Stoichiometric NO_2^--N/NH_4^+-N conversion ratios and NO_3^--$N_{produced}$/NH_4^+-$N_{consumed}$

The values of the NO_2^--N/NH_4^+-N conversion ratios observed in the batch experiments at the different incubation temperatures, COD/N ratios and COD sources are summarized in Table 2.3. At 30°C the measured conversion ratios in the control were 1.36 g-NO_2^--N/g-NH_4^+-N. This ratio is very close to the ratio reported by Strous et al., (1998) for Anammox metabolism, i.e. 1.32 g-NO_2^--N/g-NH_4^+-N.

The stoichiometric values for NO_2^--N/NH_4^+-N conversion ratios in the presence of organic matter at 30°C showed a range of 1.17-1.36 g-NO_2^--N/g-NH_4^+-N. In this regard, the conversion ratios at the initial COD/N ratio of 6 g-COD/g-N and 30 °C show values lower than those observed in the control bottles, with a minimum of 1.17 g-NO_2^--N/g-NH_4^+-N, when starch was present. On the other hand, at a COD/N ratio of 2 g-COD/g-N and 30 °C, the lowest value for the conversion ratio was 1.23 g-NO_2^--N/g-NH_4^+-N when acetate and starch were present. However, when starch was added as a single organic compound, the highest stoichiometric ratio was obtained, 1.43 g-NO_2^--N/g-NH_4^+-N. The batch tests, carried out at 22°C with a COD/N ratio of 2 g-COD/g-N, did not have significant variations in the results for almost all samples; the conversion ratios were 1.25-1.26 g-NO_2^--N/g-NH_4^+-N.

Table 2.2. Removal efficiency of NH_4^+-N, NO_2^--N and (NH_4^+ + NO_2^-)-N observed in batch experiments at different incubation temperatures, COD/N ratios and COD sources.

COD/N ratio	Sample	NH_4^+-N				NO_2^--N				(NO_2^--N+NH_4^+-N)-N			
		30°C		22°C		30°C		22°C		30°C		22°C	
		%	SD	%	SD	%	SD	%	SD	%	SD	%	SD
	Control (n=4)	57	3	71	8	100	0	100	0	76	2	84	4
2	Acetate	62	4	60	7	100	0	100	0	79	4	77	4
	Acetate + Starch	61	4	64	3	100	0	100	0	77	4	80	2
	Starch	58	4	65	3	100	0	100	0	77	3	81	2
6	Acetate	59	2	64	1	100	0	100	0	76	1	80	0
	Acetate + Starch	58	4	66	1	100	0	100	0	76	3	82	1
	Starch	55	2	66	2	100	0	100	0	73	2	82	1

SD: standard deviation

Table 2.3. NO_2^--N/NH_4^+-N conversion ratios observed in batch experiments at different incubation temperatures, COD/N ratios and COD sources.

COD/N ratio	Sample	Temperature			
		30°C		22°C	
		g-NO_2^--N/ g-NH_4^+-N	SD	g-NO_2^--N/ g-NH_4^+-N	SD
	Control (n=4)	1.36	0.09	1.19	0.09
2	Acetate	1.36	0.16	1.26	0.16
	Acetate + Starch	1.23	0.20	1.25	0.04
	Starch	1.43	0.04	1.25	0.01
6	Acetate	1.21	0.01	1.32	0.01
	Acetate + Starch	1.30	0.04	1.32	0.01
	Starch	1.17	0.33	1.36	0.08

SD: standard deviation

Higher stoichiometric values were obtained when the COD/N ratio was increased to 6 g-COD/g-N, at the same temperature, when all samples with organic matter added had conversion ratios between 1.32-1.36 g-NO_2^--N/g-NH_4^+-N.

Overall, the stoichiometric values obtained can be used as a baseline to assess the potential occurrence of denitrification by ordinary heterotrophic organisms during batch tests. In this regard, if the COD was consumed by ordinary heterotrophic organisms denitrifying over nitrite then, for instance, in accordance with Ekama and Wentzel (2008), 360 mg-COD/L would lead to about 162 mg-VSS/L (using $Y_H = 0.45$ g-VSS/g-COD), requiring around 16.2 mg-NH_4^+-N/L (assuming an N-requirement of 0.10 g-N/g-VSS) and consuming about 68.4 mg-NO_2^--N/L (based on the electron equivalents of NO_2^--N to oxygen) that would result in an NO_2^--N to NH_4^+-N ratio of about 4.22 g-NO_2^--N/g-NH_4^+-N. This ratio is considerably higher than those obtained during the tests that on average were around 1.17 - 1.43 g-NO_2^--N/g-NH_4^+-N. Results indicate that no or negligible amounts of COD were consumed by ordinary heterotrophic organisms for denitrification over NO_2^--N, implying that the N-removed via Anammox reaction was the dominant N-removal pathway.

In general, the stoichiometric NO_2^--N/NH_4^+-N conversion ratios found in the batch tests of this research are in accordance with the values reported by other authors. At 20°C, Hendrickx *et al.*, (2012) have found values between 1.11 and 1.54 g-NO_2^--N/g-NH_4^+-N during the treatment of an influent consisting of synthetic wastewater and 20% of an anaerobic effluent (on a volume basis). The anaerobic effluent was supplied from a lab-scale UASB system used for the treatment of sewage. The concentrations of NH_4^+-N and NO_2^--N in the influent were similar to those applied in this study: 30 mg/L for each nitrogen species. In the same way, in a kinetic study with Anammox cultures dominated by the species *Candidatus Brocadia caroliniensis* (flocculent sludge) and *Candidatus Brocadia fulgida* (granular sludge), Puyol *et al.*, (2013) used the obtained kinetics parameters and growing models for the estimation of the stoichiometry of the Anammox metabolism. The proposed conversion ratio of NO_2^--N/NH_4^+-N is 1.278 g-NO_2^--N/g-NH_4^+-N; this result is also close to the values of 1.25-1.26 g-NO_2^--N/g-NH_4^+-N, which were found with *Candidatus Brocadia fulgida* sludge in this study at COD/N of 2 g-COD/g-N and 22°C.

With respect to the conversion ratios at 14°C, there was no bacterial activity at this temperature (a discussion about this topic was covered in the section 2.3.1 Nitrogen removal).

Irrespective of the COD/N ratio and COD source, the NO_3^--N produced at 30°C and 22°C was below the detection limit (0.50 mg-NO_3^--N/L), except for the control, which was assessed at a COD/N ratio of 6 g-COD/g-N and at 22°C. Under these conditions, the conversion ratio of NO_3^--$N_{produced}$/NH_4^+-$N_{consumed}$ was 0.11 g-NO_3^--N/g-NH_4^+-N, which is lower than the ratio reported by Strous *et al.*, (1998) of 0.26 g-NO_3^--N/g-NH_4^+-N. The absence of nitrate production suggests a process of nitrogen removal over NO_3^--N by heterotrophs and/or possibly by the *Candidatus Brocadia fulgida* using the existing organic carbon sources. This species of Anammox bacteria has the capacity for oxidizing formate, acetate and propionate using as electron acceptor nitrite and/or nitrate (Kartal *et al.*, 2008). The coupled process of the oxidation of acetate and the nitrate reduction by *Candidatus Brocadia fulgida* has a specific activity of 0.95 µmol/min g of protein.

2.4. Conclusions

Slowly biodegradable COD (starch) and readily biodegradable COD (acetate) did not affect the metabolism of the Anammox consortia.

Observed NO_2^--N/NH_4^+-N conversion ratios show that Anammox bacteria were not impacted by denitrifying organisms in any of the performed batch tests.

References

Chernicharo, C.A.L., van Lier J.B., Noyola A., Bressani Ribeiro T., 2015. Anaerobic sewage treatment: state of the art, constraints and challenges. Reviews in Environmental Science and Bio/Technology, 14 (4), 649-679.

Dosta J., Fernández I., Vázquez-Padín J.R., Mosquera-Corral A., Campos J.L., Mata-Álvarez J., Méndez R., 2008. Short- and long-term effects of temperature on the Anammox process. Journal of Hazardous Materials, 154, 688-693.

Ekama G.A., Wentzel M.C., 2008. Nitrogen Removal. In Biological Wastewater Treatment: Principles, Modelling and Design (M. Henze, M.C.M. van Loosdrecht, G.A. Ekama and D. Brdjanovic, eds.), IWA Publishing, London, UK, pp. 87-138.

Figueroa M., Vázquez-Padín J.R., Mosquera-Corral A., Campos J.L., Méndez R., 2012. Is the CANON reactor an alternative for nitrogen removal from pre-treated swine slurry? Biochemical Engineering Journal, 65, 23-29.

Heffernan B., van Lier J.B., van der Lubbe J., 2011. Performance review of large scale up-flow anaerobic sludge blanket sewage treatment plants. Water Science and Technology, 63 (1), 100-107.

Hendrickx T.L.G., Wang Y., Kampman C., Zeeman G., Temmink H., Buisman C.J.N., 2012. Autotrophic nitrogen removal from low strength waste water at low temperature. Water Research, 46, 2187-2193.

Henze M., Comeau Y., 2008. Wastewater Characterization. In: Biological Wastewater Treatment: Principles, Modelling and Design (M. Henze, M.C.M. van Loosdrecht, G.A. Ekama and D. Brdjanovic, eds.), IWA Publishing, London, UK, pp. 33-52.

Jetten, M.S.M., Op den Camp H.J.M., Kuenen J.G., Strous M., 2010. Description of the order Brocadiales. In: Bergey's Manual of Systematic Bacteriology (N.R. Krieg, W. Ludwig, W.B. Whitman, B.P. Hedlund, B.J. Paster, J.T. Staley, N. Ward, D. Brown, A. Parte, eds.), Heidelberg, Germany: Springer, Vol. 4, pp. 596-603.

Kartal B., Rattray J., van Niftrik L., van de Vossenberg J., Schmid M.C., Webb R.I., Schouten S., Fuerst J.A., Sinninghe Damsté J.S., Jetten M.S.M., Strous M., 2007. *Candidatus* "Anammoxoglobus propionicus" a new propionate oxidizing species of anaerobic ammonium oxidizing bacteria. Systematic and Applied Microbiology, 30, 39-49.

Kartal B., van Niftrik L., Rattray J., van de Vossenberg J.L.C.M., Schmid M.C., Sinninghe Damsté J., Jetten M.S.M., Strous M., 2008. *Candidatus* 'Brocadia fulgida': an autofluorescent anaerobic ammonium oxidizing bacterium. FEMS Microbiol Ecol, 63, 46-55.

Kartal B., van Niftrik L., Keltjens J.T., Op den Camp H.J.M., Jetten M.S.M., 2012. Anammox-growth physiology, cell biology and metabolism. In: Advances in Microbial Physiology (Robert K. Poole, ed.), vol. 60, Elsevier Ltd., London, UK, pp. 211-262.

Khramenkov S.V., Kozlov M.N., Kevbrina M.V., Dorofeev A.G., Kazakova E.A., Grachev V.A., Kuznetsov B.B., Polyakov D.Y., Nikolaev Y.A., 2013. A novel bacterium carrying out anaerobic ammonium oxidation in a reactor for biological treatment of the filtrate of wastewater fermented sludge. Microbiology, 82 (5), 628-636.

Liu T., Li D., Zeng H., Chang X., Zhang J., 2013. Microbial characteristics of a CANON reactor during the start-up period seeding conventional activated sludge. Water Science and Technology, 67 (3), 635-643.

Lotti T., van der Star W.R.L., Kleerebezem R., Lubello C., van Loosdrecht M.C.M., 2012. The effect of nitrite inhibition on the anammox process. Water Research, 46, 2559-2569.

Lotti T., 2013. Identification of Anammox species in Dockhaven wastewater treatment plant. Personal communication between T. Lotti and J. A. Sánchez Guillén (not published), Department of Biotechnology, Delft University of Technology, Julianalaan 67, Delft 2628 BC, the Netherlands.

NEN, 1983. Photometric determination of ammonia in Dutch system. In: Nederlandse Normen (Dutch Standards), (International Organization for Standardization, ed.) NEN 6472, Dutch Institute of Normalization, Delft, the Netherlands.

Oshiki M., Shimokawa M., Fujii N., Satoh H., Okabe S., 2011. Physiological characteristics of the anaerobic ammonium-oxidizing bacterium 'Candidatus Brocadia sinica'. Microbiology, 157, 1706-1713.

Park H., Rosenthal A., Jezek R., Ramalingam K., Fillos J., Chandran K., 2010. Impact of inocula and growth mode on the molecular microbial ecology of anaerobic ammonia oxidation (anammox) bioreactor communities. Water Research, 44, 5005-5013.

Procópio Pontes P. and Chernicharo C.A.L., 2011. Characterization and removal of specific organic constituents in an UASB-trickling-filter system treating domestic wastewater. Environmental Technology, 32 (3), 281-287.

Puyol D., Carvajal-Arroyo J.M., Garcia B., Sierra-Alvarez R., Field J.A., 2013. Kinetic characterization of *Brocadia* spp.-dominated anammox cultures. Bioresource Technology, 139, 94-100.

Rattray J.E., van de Vossenberg J., Jaeschke A., Hopmans E.C., Wakeham S.G., Lavik G., Kuypers M.M.M., Strous M., Jetten M.S.M., Schouten S., Sinninghe Damsté J.S., 2010. Impact of temperature on ladderane lipid distribution in Anammox bacteria. Applied and Environmental Microbiology, 76 (5), 1596-1603.

Seitzinger S.P., Harrison J.A. 2008 Land-based nitrogen sources and their delivery to coastal systems. In: Nitrogen in the Marine Environment (D. Capone, D.A. Bronk, M.R. Mullholland, E. Carpenter, eds.), 2nd edn, Academic Press, New York, pp. 469-510.

Standard Methods for the Examination of Water and Wastewater, 2012, 22 ed. American Public Health Association/American Water Works Association/Water Environment Federation, Washington DC, USA.

Strous M., Heijnen J.J., Kuenen J.G., Jetten M.S.M., 1998. The sequencing batch reactor as a powerful tool for the study of slowly growing anaerobic ammonium- oxidizing microorganisms. Appl. Microbiol, 50 (5), 589-596.

Strous M., Kuenen J.G., Jetten M.S.M., 1999. Key physiology of anaerobic ammonium oxidation. Applied and Environmental Microbiology, 65 (7), 3248-3250.

van de Graaf A.A., de Brujin P., Robertson L.A., Jetten M.S.M., Kuenen J.G., 1997. Metabolic pathway of anaerobic ammonium oxidation on the basis of ^{15}N studies in a fluidized bed reactor. Microbiology, 143, 2415-2421.

Veys P., Vandeweyer H., Audenaert W., Monballiu A., Dejans P., Jooken E., Dumoulin A., Meesschaert B.D., van Hulle S.W.H., 2010. Performance analysis and optimization of autotrophic nitrogen removal in different reactor configurations: a modelling study. Environmental Technology, 31 (12), 1311-1324.

Winkler M.K.H., Kleerebezem R., van Loosdrecht M.C.M., 2012. Integration of anammox into the aerobic granular sludge process for main stream wastewater treatment at ambient temperatures. Water Research, 46, 136-144.

Long-term performance of the Anammox process under low nitrogen sludge loading rate and moderate to low temperature

Contents

This chapter is based on:

Sánchez Guillén J.A., Lopez Vazquez C.M., de Oliveira Cruz L.M., Brdjanovic D., van Lier J.B., 2016. Long-term performance of the Anammox process under low nitrogen sludge loading rate and moderate to low temperature. Biochemical Engineering Journal, 110, 95-106.

Abstract

The Anammox process was studied during 1048 days in a Sequencing Batch Reactor (SBR) fed with a mineral medium, under a Nitrogen Sludge Loading Rate (NSLR) less than 0.080 g-N/g-VSS·d and a temperature range of 30.5±0.5 °C - 13.2±0.3 °C. Anammox granular biomass from a full-scale plant treating sludge reject water was used as inoculum; this plant had a NSLR of 0.238 g-N/g-TSS·d at 34±2.5 °C. The research was divided in four phases according to the NSLRs and temperatures applied. In order to assess the long-term influence of a NSLR lower than the NSLR capacity of the biomass and the diminishing of temperature on the Anammox process, the total nitrogen removal efficiency, specific activity, dynamics in granule size distribution, biomass concentration and microbial population variations were analyzed. The results provide useful information for the potential operation of an Anammox SBR for sewage (post) treatment using an inoculum with a high NSLR.

3.1. Introduction

Since about 2001, the Anammox process has been applied at full scale for the removal of Dissolved Inorganic Nitrogen (DIN) in landfill leachate, industrial wastewater and sludge reject water in municipal sewage treatment plants. With respect to sewage, the elimination of DIN via Anammox process has been focused on sludge reject water coming from sludge handling facilities, i.e. side-stream treatment lines. This so-called 'sludge reject water' is characterized by its low chemical oxygen demand (COD) to nitrogen ratio, i.e. usually a COD/N < 2 and high DIN concentration and temperature, e.g. 1000 mg-N/L and 34^{0}C, correspondingly (Lackner *et al.*, 2014).

Wastewaters were initially treated by two-step Anammox systems, which consist of partial nitritation followed by autotrophic denitrification using ammonium as electron donor and nitrite as electron acceptor in two consecutives bioreactors, respectively. Hereafter one-step Anammox schemes were developed, where partial nitritation and Anammox conversion occurs simultaneously in the same reactor. Efficient upstream organic carbon removal, effective biomass retention capacity, low temperatures, handling of low DIN concentrations and consequently a low Nitrogen Sludge Loading Rate (NSLR), as well as high nitrogen removal efficiency, have been identified as the major challenges for the application of the Anammox process in the main stream of sewage treatment plants (Lotti *et al.*, 2014a; Lotti *et al.*, 2015; Sánchez Guillén *et al.*, 2015a; Sánchez Guillén *et al.*, 2015b).

Regarding upstream organic carbon removal in sewage and downstream elimination of DIN via Anammox metabolism, some authors have proposed the combination of methanogenesis and Anammox processes in a single integrated system by the coupling of Upflow Anaerobic Sludge Blanket (UASB) reactor and Anammox reactor (one or two-steps) (Jetten et al., 1999; Gujer, 2010; Lotti *et al.*, 2014a; Sánchez Guillén *et al.*, 2014; Chernicharo *et al.*, 2015). This treatment scheme would provide important benefits such as energy efficiency and distinct drop in sludge production. Some trials for the coupling of UASB and Anammox reactors have been reported by Malamis *et al.* (2013) and Malovanyy *et al.* (2015); the treatment scheme suggested by Malamis *et al.* (2013) was a UASB reactor followed by an Anammox one-step Sequencing Batch Reactor (SBR).

Lackner *et al.* (2014) highlighted the use of full-scale Anammox SBR reactors through a survey of 100 full-scale Anammox treatment plants operating worldwide. This research group has identified that more than 50% of full-scale Anammox facilities correspond to SBR technology. Furthermore, in terms of average nitrogen load per plant, they have found that the granular systems treat the majority of the nitrogen content and 75% of the plants are installed for nitrogen removal from sludge reject water in sewage treatment.

In general, Anammox SBRs are efficient biomass retention systems (Strous *et al.*, 1998). The sludge loading rate is an important parameter for the design and operation of these bioreactors (van Loosdrecht, 2008). In eight surveyed full-scale Anammox SBRs treating sewage sludge reject water, the operational Nitrogen Sludge Loading Rate (NSLR) was in the range of 71-155 g-N/kg-TSS·d (Lackner *et al.*, 2014). In the scenario of a UASB-Anammox SBR system for DIN removal in the main sewage stream, the inoculation of the Anammox SBR with Anammox granules coming from a full-scale Anammox treatment plant that operates with high NSLR would represent a new challenge. The proposed use may lead to quality loss of the inoculum because its NSLR capacity is distinctly higher compared to the NSLR applied in the UASB-Anammox SBR system, possibly leading to impairment of the Anammox process.

Lotti *et al.* (2014a) have studied the treatment of influents with low nitrogen strength at low temperatures. These researchers have achieved a total nitrogen removal of 39% during the treatment of synthetic sewage at 10^{0}C and 60 mg-NH_4^+-N/L by means of a lab-scale gas-lift Anammox one-step SBR. Similarly, Gilbert *et al.* (2014) used an Anammox one-step moving bed biofilm reactor (MBBR) for treating a synthetic wastewater with a low DIN concentration, i.e. 50 mg-NH_4^+-N/L at temperatures from 20^{0}C to 10^{0}C. The drop in temperature caused the nitrogen removal rate to diminish by approximately 63%, requiring doubling the hydraulic retention time (HRT) in order to keep the nitrogen concentration less than 8 mg-NH_4^+-N/L in the effluent.

Thus far, the long-term performance of the Anammox process under low NSLR has not been investigated before. Authors working on the topic of UASB-Anammox systems treating sewage, have not take into account the influence of the very low NSLR on the long-term stability of Anammox processes. For instance, Malamis *et al.* (2013) used an inoculum with a dominant heterotrophic activity (denitrifiers) and a very low Anammox specific activity, i.e. 0.02 kg-N/kg-VSS·d at 30^{0}C. During the start-up of the lab-scale Anammox SBR, the NSLR applied was eight times higher than Anammox specific activity of the inoculum, i.e. 0.17 kg-N/kg-VSS·d. Furthermore, in the last phase of their research the NSLR applied and the Anammox specific activity had similar values, i.e. 0.06 kg-N/kg-VSS·d and 0.05 kg-N/kg-VSS·d, correspondingly. Therefore the removal capacity of the inoculum and the Anammox biomass during the research of Malamis *et al.* (2013) did not experience any substantial restriction in terms of nitrogen loading rate availability. Malovanyy *et al.* (2015) coupled a UASB reactor and an Anammox one-step MBBR that was used during three years for sludge reject water treatment. During the first period of the research, the MBBR treated a mixture of sludge reject water and UASB effluent (20°C) obtained from sewage treatment. The volumetric ratio of the mixture of the UASB effluent and the sludge reject water was decreased step-wise during 157 days until the MBBR only treated the UASB effluent. Consequently, the NSLR in the MBBR was diminished from 0.062 g-N/g-VS·d to 0.009 g-N/g-VS·d. The specific Anammox activity was around five times higher than the nitrogen load applied and part of the Anammox bacteria died-off because there was not sufficient substrate available. At the end of their research, the DIN removal achieved was 34% from a UASB effluent with a content of 34 mg-NH_4^+-N/L.

Considering the potential effects of the NSLR on the performance of the Anammox process during main stream sewage treatment, the lack of information regarding this topic and the widespread use of the SBR full-scale systems, the main objective of this research is to perform a long-term study of the Anammox process, in an SBR under low NSLR and moderate to low temperature. To achieve this purpose, a mineral medium was used that simulates a UASB effluent after partial nitritation. The medium did not include organic carbon, in order to minimize the interference of denitrifiers during the research. The mineral medium was treated by a granular Anammox SBR at 30.5 ± 0.5-13.2 ± 0.3^{o}C during 1048 days.

3.2. Materials and methods

3.2.1. Configuration and operation of the SBR

This study was carried out using an Anammox SBR with a working volume of 10 L. The inoculum was obtained from an Anammox upflow granular bed reactor (Dokhaven-Sluisjesdijk wastewater treatment plant in Rotterdam, the Netherlands). This bioreactor is used for the treatment of the effluent coming from a partial nitritation reactor, i.e. a SHARON-ANAMMOX system that treats sewage sludge reject water (van der Star *et al.*, 2007).

The operation of the Anammox SBR started with an adaptation period of 450 days. During this time, the performance of the Anammox SBR was recorded for assessing the granular biomass acclimatization to the new operational and environmental conditions, i.e. temperature, NLR, NSLR, NO_2^--N+NH_4^+-N influent concentration, pH, feeding and mixing regime (Table 3.1). The adaptation period was followed by four phases defined mainly by the temperature applied (Table 3.2).

The SBR was equipped with three vertical baffles and a vertical shaft having two Rushton impellers (4 blades) used for mixing. Temperature control in the SBR was achieved by an internal serpentine (inner heat exchanger) connected to a digital waterbath. The SBR was operated and controlled automatically through an Applikon BioController ADI 1030 (Applikon; Schiedam, the Netherlands). Applikon sensors for recording temperature, pH and dissolved oxygen (DO) with an accuracy of ±0.1^{0}C, ±0.01 pH and ±0.1%, respectively, were utilized; the data from these parameters was stored online via the BioXpert software (Applikon; Schiedam, the Netherlands). Temperature and pH were varied as is described in Table 3.2; pH was adjusted by dosing 0.4 M NaOH and 0.4 M HCl.

The SBR was operated in cycles of 6 h; each cycle was composed by 5 phases (Figure 3.1). The cycle began with an initial volume of 7.5 L, mixing at 180 rpm and flushing of nitrogen gas was applied during 10 min by an "L-shaped" sparger submerged in the liquid phase to remove the DO; a maximum value of 0.2% of DO in the bulk liquid phase was maintained during the entire operation of the SBR. In the following 300 min, the mixing continued and the SBR was fed by 2.5 L of synthetic substrate that simulates a DIN rich effluent (around 50 mg of NH_4^+-N/L + 50 mg of NO_2^--N/L) from a UASB system treating sewage. When the feeding was finalized, a period of 20 min of mixing was applied to remove the produced nitrogen gas from the Anammox granules facilitating the biomass retention in the bioreactor (Dapena-Mora *et al.*, 2004a). Afterwards, the mixing was switched off and 15 min of settling and 15 min for withdrawal of 2.5 L of effluent were applied; thus, the volumetric exchange ratio per cycle was 25%. The HRT was 1 d and the Nitrogen Loading Rate (NLR) was around 0.100 kg-N/m^3·d, except during the last days of the experiment (Table 3.2). No waste of sludge was applied; therefore the wasted solids were only those in the effluent. For the entire period of the research, the sludge retention time (SRT) varied according to the concentration of solids in the SBR and the effluent. The SRT was initially estimated at 957 days, dropping to 525 days at the end of the study, i.e. day 1048. In order to avoid air intrusion in the SBR, the headspace of the reactor was connected to a water-lock. Gas bubble formation was observed in the water-lock caused by the nitrogen production via Anammox bacteria.

3.2.2. Mineral medium

The mineral medium was prepared using two solutions, i.e. ammonium and nitrite rich feeds, that were mixed with demineralized water. The total influent concentration of NH_4^+-N+NO_2^--N was approximately 100 mg-N/L, except during the last part of Phase IV, which was around 32 mg-N/L. The feeding ratio NO_2^--N/NH_4^+-N for each phase is shown in Table 3.2. The ammonium and nitrite rich feeds were prepared according to van de Graaf *et al.* (1996) with some modifications. The composition of ammonium and nitrite feeds, per 1 liter of demineralized water, was (i) ammonium feed: 2.9828 g NH_4Cl; 4.6875 g $CaCl_2 \cdot 2H_2O$; 0.3906 g KH_2PO_4; 0.77 g $MgSO_4 \cdot 7H_2O$; (ii) nitrite feed: 3.8504 g $NaNO_2$; 19.531 g $KHCO_3$; 0.1786 g NaEDTA; 0.1786 g $FeSO_4 \cdot 7H_2O$ and 1.25 mL of trace element solution. Per liter of demineralized water, the trace element solution was composed by: 0.05 g $NaWO_4 \cdot 2H_2O$; 0.25 g $CuSO_4 \cdot 5H_2O$; 0.22 g $Na_2MoO_4 \cdot 2H_2O$; 0.43 g $ZnSO_4 \cdot 7H_2O$; 0.24 g $CoCl_2 \cdot 6H_2O$; 0.99 g $MnCl_2 \cdot 4H_2O$; 0.19 g $NiCl_2 \cdot 6H_2O$; 0.1076 g Na_2SeO_4; 0.014 g H_3BO_3; 15 g Mg EDTA.

3.2.3. Analytical methods

The size of the Anammox granules (average equivalent diameter) was measured by imaging particle analysis. These tests were performed using a microscope Leica Microsystems M205 FA (software version Qwin V3.5.1., calibration factor 4.45, magnification 13.0; Leica Microsystems Ltd, the Netherlands). The procedures described in Methods for the Examination of Water and Wastewater (2012) were used for the analysis of the Mixed Liquor Total Suspended Solids (MLTSS), the Mixed Liquor Volatile Suspended Solids (MLVSS), and nitrite-nitrogen (NO_2^--N). The optimum depth in the SBR for obtaining a representative sample for MLSS and MLVSS analyses was determined at 12cm (measured from the SBR's bottom). This depth was estimated using the results from the analyses of the MLSS and MLVSS at depths of 0, 8, 16 and 24 cm (bottom to top; the total working volume depth was 26 cm) and applying the discrete volumes method for the calculations (See Appendix 3A). The MLTSS and MLVSS in the effluent were analyzed from a volume of 10 L that was collected over 1 HRT, i.e. 4 consecutive cycles. The 10 L of effluent was well mixed during 25 minutes and samples of 100 mL were taken for the tests. The standard NEN 6472 (NEN, 1983) was utilized for the measurement of ammonia-nitrogen (NH_4^+-N). The nitrate-nitrogen (NO_3^--N) was tested by the standard ISO 7890-1:1986 (ISO 7890/1, 1986). All samples used for the determination of NH_4^+-N, NO_2^--N and NO_3^--N were previously filtered by 0.45 μm Ion Chromatography (IC) Acrodisc® Syringe Filters.

Table 3.1. Operational and environmental conditions of the bioreactor used as source of Anammox inoculum (Lotti *et al.*, 2012; Lackner *et al.*, 2014) and the Anammox SBR.

Feature	**Anammox upflow granular bed reactor**	**Anammox SBR**
Feeding	Continuous feeding	There was no feeding by an hour in each cycle (see Figure 3.1)
Mixing	Hydraulic mixing	Mechanical mixing by a vertical shaft having two Rushton impellers (4 blades). Mixing was stopped during the last 30 minutes of every cycle (phases of settling and draw)
Temperature	34 ± 2.5 °C	30.5 ± 0.5 °C
pH	7.2 ± 0.4	7.89 ± 0.01
Influent NO_2^--N+NH_4^+-N	1000 mg-N/L	100 mg-N/L
NLR	7.03 kg-N/m^3·d	0.098 ± 0.014 kg-N/m^3·d
NSLR	0.238 g-N/g-TSS·d	0.020 g-N/g-VSS·d

Table 3.2. Operational phases of the Anammox SBR, nitrogen removal efficiency and stoichiometric conversion ratios.

Phase	T (°C)	Period (days)	pH	NLR (kg N/m^3·d)	Feeding ratio NO_2^--N/NH_4^+-N	Total N influent (mg N/L)	Total N removal (%)		Stoichiometric conversion ratios NO_2^--N/NH_4^+-N		NO_3^--N/NH_4^+-N	
							Average	Last day of phase	Average	Last day of phase	Average	Last day of phase
Adaptation	30.5 ± 0.5	1-450	7.89 ± 0.01	0.098 ± 0.014	0.92 ± 0.10	103 ± 14	76 ± 12	73	1.30 ± 0.18	1.32	0.27 ± 0.18	0.31
I	30.5 ± 0.5	450-743	7.89 ± 0.01	0.114 ± 0.033	0.99 ± 0.17	116 ± 35	76 ± 7	71	1.35 ± 0.12	1.23	0.29 ± 0.07	0.39
II	25.5 ± 0.5	743-823	7.89 ± 0.01	0.100 ± 0.000	0.91 ± 0.00	100 ± 0	72 ± 2	73	1.28 ± 0.03	1.26	0.34 ± 0.05	0.31
III	18.0 ± 0.1	823-945	7.89 ± 0.01	0.103 ± 0.009	1.11 ± 0.07	104 ± 10	86 ± 5	81	1.22 ± 0.04	1.25	0.22 ± 0.08	0.24
IV	16.8 ± 0.1	945-952	7.89 ± 0.01	0.098 ± 0.001	0.94 ± 0.10	98 ± 1	71 ± 6	75	1.27 ± 0.05	1.32	0.30 ± 0.05	0.33
	16.0 ± 0.3	952-959	8.00 ± 0.04[a]	0.100 ± 0.001	1.04 ± 0.06	103 ± 6	63 ± 10	70	1.38 ± 0.13	1.47	0.38 ± 0.04	0.35
	14.7 ± 0.3	959-966	7.94 ± 0.09[a]	0.090 ± 0.008	0.96 ± 0.10	94 ± 8	62 ± 4	67	1.73 ± 0.19	1.47	0.42 ± 0.04	0.36
	13.8 ± 0.2	966-973	7.89 ± 0.05[a]	0.095 ± 0.013	0.93 ± 0.10	99 ± 7	39 ± 13	30	1.31 ± 0.15	1.20	0.51 ± 0.01	0.51
	13.2 ± 0.3	973-1048	8.25 ± 0.00[b]	0.032 ± 0.004[b]	0.98 ± 0.23[b]	32 ± 3[b]	91 ± 4[b]	93[c]	1.13 ± 0.19[b]	1.12[c]	0.02 ± 0.02[b]	0.01[c,d]

[a]No pH control. [b]From day 1008 to 1016. [c]Day 1016. [d]On day 1024 the ratio increased to 0.15 NO_3^--$N_{produced}$/NH_4^+-$N_{consumed}$.

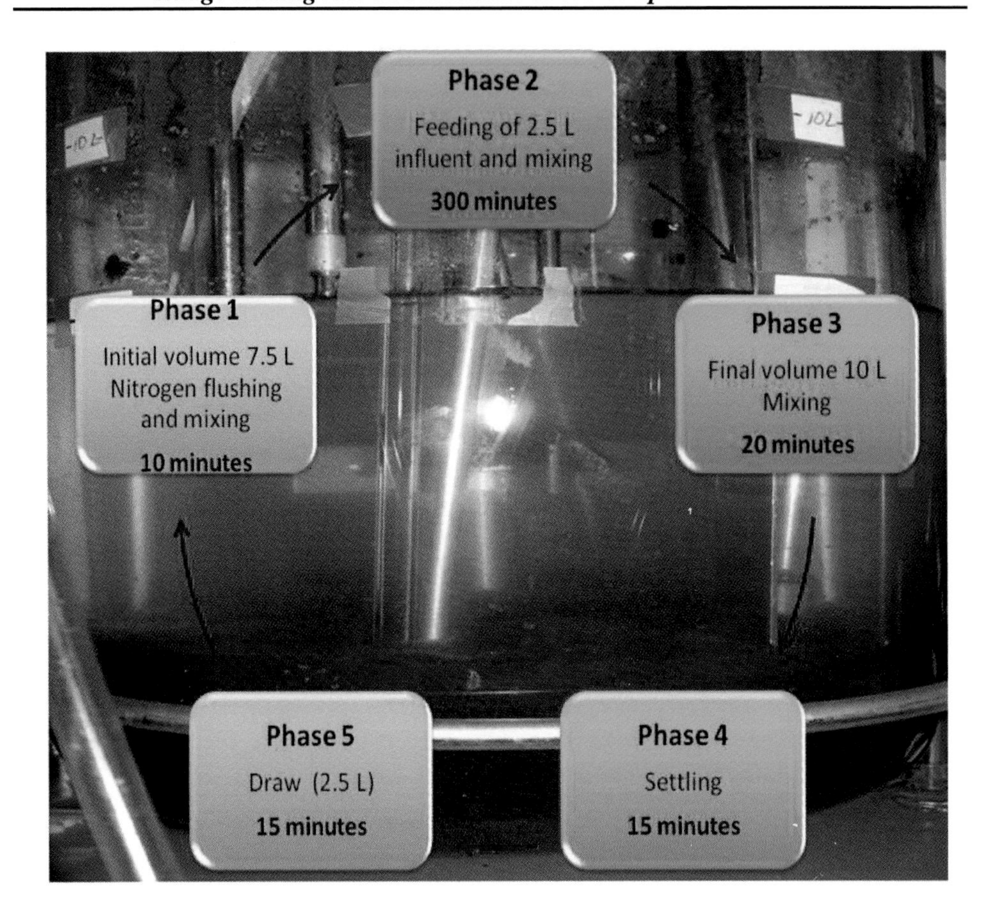

Figure 3.1. Cycle of the Anammox Sequencing Batch Reactor (SBR).

3.2.4. Anammox in situ activity tests

For each experimental phase, the specific activity of the biomass of the SBR was measured *in situ*, i.e. instead of the normal feeding of the cycle, 2.5 liters of a pulse feeding containing NH_4^+-N and NO_2^--N in an approximate ratio 1:1 and with an influent nitrogen concentration of 100 mg-N/L was added at once to the SBR. Every 8-10 minutes samples were collected from the SBR and the concentration of NH_4^+-N, NO_2^--N and NO_3^--N were measured until the NO_2^--N was totally consumed (a period of about 80 minutes). From these results the graphs of the consumption of the NH_4^+-N and NO_2^--N *vs.* time were obtained and based on the biomass concentration (volatile suspended solids) inside the SBR during the test, the specific activity was calculated in terms of g-$(NO_2^- + NH_4^+)$-$N_{consumed}$/(g-VSS·d).

3.2.5. Identification of microbial populations

Fluorescence *in Situ* Hybridization (FISH) method was performed to identify the microbial population in the SBR. Biomass samples were taken from a depth of 12 cm (from bottom to top) for the analysis. These samples were fixed in paraformaldehyde (4%) and the procedure described by Schmid *et al.* (2000) for hybridizations with fluorescent probes was performed. Epifluorescence was used for detection of the ammonia-oxidizing organisms (AOO), nitrite-oxidizing organisms (NOO), Anammox bacteria and 4', 6'-diamidino-2-phemylindol (DAPI) as general DNA stain. The fluorophores Cy3 or FLUOS (Biomers.net, Germany) were utilized for labeling all gene probes. The FISH images were obtained via the standard software package (Version 1.2) of the epifluorescence microscope OLYMPUS BX51-CMOS (color camera for light microscopy, Japan).

3.3. Results and Discussion

Removal of total nitrogen over the entire period of research is shown in Figure 3.2.

3.3.1. Adaptation period

The influent of the SBR was supplied with a ratio of substrates equal to 0.92 ± 0.10 mg-NO_2^--N/mg-NH_4^+-N with the aim of simulating a stream coming from a partial nitritation reactor, e.g. a SHARON reactor. The applied NLR was 0.098 ± 0.014 kg-N/m^3·d. The NLR is a parameter expressing the nitrogen loading to the system. The NLR can be linked to the biomass content, which then expresses the nitrogen sludge loading rate (NSLR) or the food to microorganism (F/M) ratio, e.g. g-N/g-VSS·d. In a biological reactor with a stable biomass concentration and stable operational conditions, the NSLR and the sludge nitrogen removal capacity determine the specific nitrogen removal rate of the system.

Regarding the specific nitrogen removal rate of the system, the NSLR might comply, exceed or does not cover the nitrogen removal capacity of the biomass. The inoculum for the Anammox SBR reactor came from an Anammox upflow granular bed reactor with an NSLR of 0.238 g-N/g-TSS·d (Lackner *et al.*, 2014). However, the NSLR in the Anammox SBR was 0.020 g-N/g-VSS·d. This value is around 12 times less compared to the NSLR of the Anammox upflow granular bed reactor and approximately 23 times less than the maximum specific activity of the inoculum that is 0.458 g-N_2-N/g-VSS·d (Lotti *et al.*, 2012). Hence, in this scenario the nitrogen removal capacity of the microorganism in the SBR was deliberately not satisfied in order to evaluate the adaptation of the granular biomass to such condition. Despite the low NSLR applied, the total nitrogen removal achieved by the SBR at the end of the adaptation period was 73%. The stoichiometric nitrogen removal ratio and the nitrate production ratio were 1.32 g-NO_2^--N/g-NH_4^+-N and 0.31 g-NO_3^--$N_{produced}$/g-NH_4^+-$N_{consumed}$, respectively; these ratios are in accordance with the Anammox metabolism reported in other granular SBR reactors (Strous *et al.*, 1998).

The imposed NSLR may also impact the morphological structure of the microbial aggregates, i.e. granules. A decrease in the loading rate may results in deterioration of the integrity of granules. Alphenaar (1994) and McKeown *et al.* (2009) observed this phenomenon in anaerobic sludge bed systems, i.e. an upflow anaerobic sludge blanket (UASB) reactor and an expanded granular sludge bed-anaerobic filter (EGSB-AF) hybrid reactor, respectively. The mass transfer limitation was an important factor in the granule disintegration.

Several authors have modelled Anammox-based granular bed systems assuming a well-mixed reactor and neglecting external mass transfer limitations (Volcke *et al.*, 2012; Mozumder *et al.*, 2014). However, it is important to consider the impact of the imposed NSLR on the reactor's performance because there are internal mass transfer limitations in the Anammox granules (Ni *et al.*, 2009). After a drastic drop in the applied NSLR in the Anammox SBR, the internal mass transfer limitations might induce starvation of the biomass, which would lead to granule disintegration. Granule fragmentation might cause, to some extent, a washout of suspended biomass with the effluent and the subsequent diminishing of the biomass concentration in the Anammox SBR reactor with time.

The results obtained from the Anammox SBR on day 84, showed that the biomass concentration inside this reactor was 4000 mg-VSS/L, but 163 days later, about 48% reduction of the VSS was registered in the reactor content. Hereafter, the biomass concentration apparently stabilised in the SBR, being 2120 mg-VSS/L 202 days later (Figure 3.3.A). The ratio VSS/TSS of the SBR (VSS_{SBR}/TSS_{SBR}) experienced also a declining behaviour; passing from 0.680 ± 0.030 at day 84 to 0.560 ± 0.010 at day 449 which indicates an accumulation of inert material inside the reactor. This behaviour is also confirmed by the results of the VSS/TSS ratio in the effluent (VSS_{eff}/TSS_{eff}) showing a value of 0.900 ± 0.010, indicating that biomass was the predominant constituent of the suspended solids in the effluent.

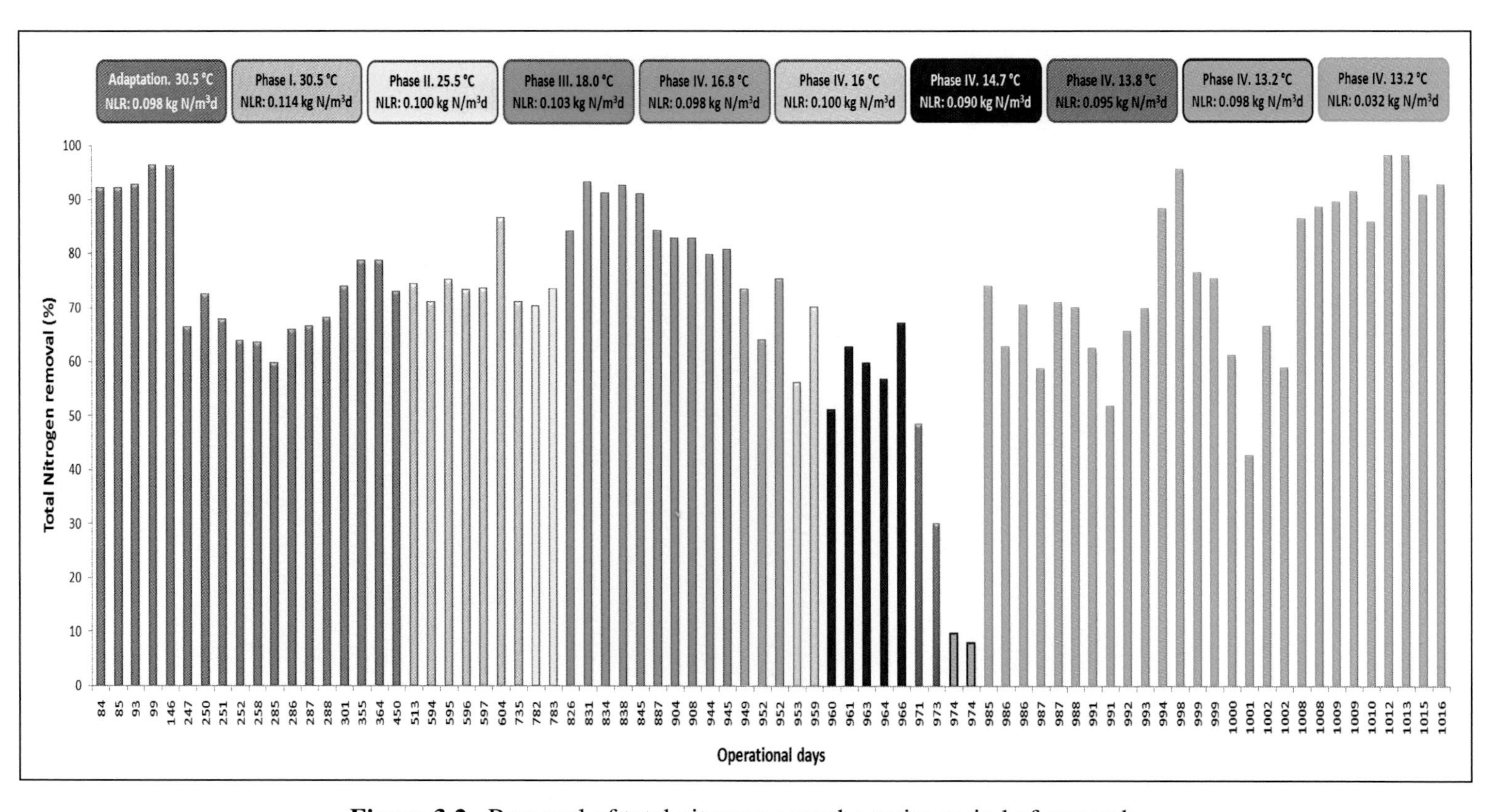

Figure 3.2. Removal of total nitrogen over the entire period of research.

Consequently, the biomass concentration in the effluent at day 84 was 4.1 ± 0.1 mg-VSS/L, but on day 247 it decreased to 2.9 ± 0.3 mg-VSS/L. A similar value was measured at the end of the adaptation period, i.e. 3.2 ± 0.4 mg-VSS/L. The average biomass loss in the effluent for the entire period was very low, i.e. 31 ± 6 mg-VSS/d (Figure 3.3.B).

The integrity of the granular biomass is also affected by the shear stress. The shear stress on granular biomass can be caused by hydraulic mixing (Lotti *et al.*, 2014a) or mechanical mixing, leading to granule erosion and breakage. Some authors have studied the influence of shear stress on the Anammox granules. However, in these studies the NSLR suffered variations and its impact on the granule integrity was not taken into consideration. For instance, Arrojo *et al.* (2006) conducted a research regarding the effects of mechanical stress on Anammox granules in an SBR which was previously used by Dapena-Mora *et al.* (2004b) for assessing the stability of the Anammox process.

During the research conducted by Dapena-Mora *et al.* (2004b), the SBR was operated at 35^{0}C with an NSLR of 0.280 g-N/g-VSS·d, mechanical mixing at 70 rpm and a controlled pH of 7.8-8.0. The new operational conditions of the SBR at the starting point of the research of Arrojo *et al.* (2006) indicated an operating temperature of 30 ± 1^{0}C, an NSLR in the range of 0.167-0.187 g-N/g-VSS·d, which is roughly 2 times less than the NSLR applied by Dapena-Mora *et al.* (2004b), mechanical mixing at 60 rpm and a no pH control; pH values were in the range of 7.5-8.0.

Although the initial mixing in the SBR was set to 60 rpm (specific input power of 0.003 kW/m^{3}), during this phase the VSS content of the SBR decreased from approximately 2 g-VSS/L (day 20) to 1.650 g-VSS/L (day 80) and the VSS concentration of the effluent had an increment of around 0.080 g-VSS/L. Arrojo *et al.* (2006) explained the increase in the effluent VSS concentration as a consequence of a shock in the pH (not reported value). However, taking into account the decrease in the average diameter of the granular biomass in the SBR (from about 1.05 to 0.75 mm) for the same period, it could be possible that this change in the granule size is also related with the decrease in the VSS content in the SBR. Considering that there was not any significant change in the original mixing conditions (Dapena-Mora *et al.*, 2004b), the diminishing size of the granular biomass, the decrease in the SBR VSS concentration and the increase in the effluent VSS concentration might have been caused by the reduction in the NSLR applied by Arrojo *et al.* (2006). On the other hand, Arrojo *et al.* (2006) demonstrated that the use of a high stirring speed, e.g. 250

rpm (equivalent to a specific input power of 0.23 kW/m^3) produced a substantial shear stress on the granules, reducing the size of the granular aggregates and increasing the biomass washout.

Similarly, Durán *et al.* (2014) attributed the disintegration of the granules and the loss of biomass, from an Anammox SBR, to the mechanical mixing regime applied, i.e. 120 rpm during 280 days. This Anammox SBR was inoculated with Anammox granules from a pilot plant and operated at 24°C. The initial conditions of the SBR in the research of Durán *et al.* (2014) included a specific Anammox activity of 0.21 g-N_2-N/g-VSS·d, an NLR of 0.10 ± 0.01 g-N/L·d and an NSLR of 0.02 g-N/g-VSS·d which is 10 times less than the assessed specific Anammox activity. During the first 40 days of operation, the NLR was increased to 0.20 ± 0.01 g-N/L·d, but for the rest of that research it was stabilised to 0.10 ± 0.01 g-N/L·d. Since the start-up of the SBR of Durán *et al.* (2014), the VSS concentration of the reactor was decreasing until it reached a 59% of the initial value, at day 125. Nevertheless, from this day and during the rest of the experimental period, the biomass concentration remained constant.

The results obtained at a low NSLR from the research of Durán *et al.* (2014) and from our study with the Anammox SBR in the adaptation period, share some similarities in terms of a decrease in the VSS concentration, a stable VSS concentration for a certain period of time and a final NSLR determined by the reduction of the biomass concentration in the reactors. In the SBR of Durán *et al.* (2014) and in our Anammox SBR, the VSS decrease was accounted to 41% and 48%; the VSS concentration was constant for 155 and 202 days and the final NSLR was 0.04 and 0.05 g-N/g-VSS·d, respectively.

From the previous discussion, it may be hypothesized that the changes in the size of the Anammox granules and in the biomass concentration of a granular SBR reactor under a mechanical mixing regime are not only influenced by the shear stress but also by the diminishing of the applied NSLR. However, it is difficult to distinguish separately to what extent the decrease in the NSLR and the mechanical mixing contributed to the granule disintegration and biomass washout during the adaptation period of the Anammox SBR. Further studies are necessary to elucidate the scope of each mechanism.

3.3.2. Phase I

The temperature was kept at 30.5 ± 0.5°C during Phase I. At day 62 of this phase (operational day 512), the VSS concentration of the SBR was comparable to the value at the end of the adaptation phase, around 2000 mg-VSS/L (Figure 3.3.A). But, the VSS_{SBR}/TSS_{SBR} ratio was 0.49 ± 0.03, which was 12 % less than the obtained ratio at the final stage of the adaptation phase. This result indicates that the trend of accumulation of inert material in the SBR was still ongoing. The specific Anammox activity was 0.329 g-(NO_2^--N+NH_4^+)-$N_{consumed}$/(g-VSS·d) and, at the beginning of Phase I, the sizes of the Anammox granules at depths of 0, 12 and 20 cm (measured from the bottom to the top of the Anammox SBR) were very similar. The average diameter of the granules was in the range of 390-412 μm (Figure 3.4.A-B), which denotes a decrease of 63-65% in the initial inoculum size.

The diminishing of the size of the granules might be caused by the NSLR applied and the mechanical mixing during the adaptation phase as it was discussed in section 3.1. Figure 3.5 shows the NSLR capacity of the Anammox SBR reactor *vs.* the NSLR applied. The NSLR capacity of the Anammox SBR reactor was calculated based on its operational conditions, i.e. feeding regime, HRT, the volumetric activity obtained from the activity tests performed at the different temperatures of each phase (Figure 3.4.A) and the VSS concentration in the SBR (Figure 3.3.A). For example, based on the duration of feeding, number of cycles per day and the HRT, that is 5 h, 4 cycles and 1 d, respectively and with a volumetric activity of 0.660 g-(NO_2^-+NH_4^+)-N/L·d at 30.5 ± 0.5°C, the maximum nitrogen influent concentration that the SBR was able to treat was 550 mg-(NO_2^-+NH_4^+)-N/L. This is in agreement with an NLR capacity of 0.550 g-N/L·d and a biomass concentration of 1.940 g-VSS/L, which characterizes an NSLR capacity of 0.284 g-N/g-VSS·d. However, the NSLR utilized was only 21% of the NSLR capacity that the SBR could have handled (Figure 3.5).

At day 589 and during the following 8 days, the pumps that were feeding the substrate were not functioning properly. As a consequence of this unforeseen situation, the total nitrogen concentration in the influent was higher than 100 mg-N/L, i.e. 160 mg-N/L with a feeding ratio of 0.90 mg-NO_2^--N/mg-NH_4^+-N. Consequently, the NLR increased to 0.160 kg-N/m^3·d, during this event, which agrees with an increment of 63% in the NSLR. For the duration of this event, the performance of the Anammox SBR demonstrated that the total nitrogen removal capacity of the system and the stoichiometry were similar to the results of the last days of the adaptation period.

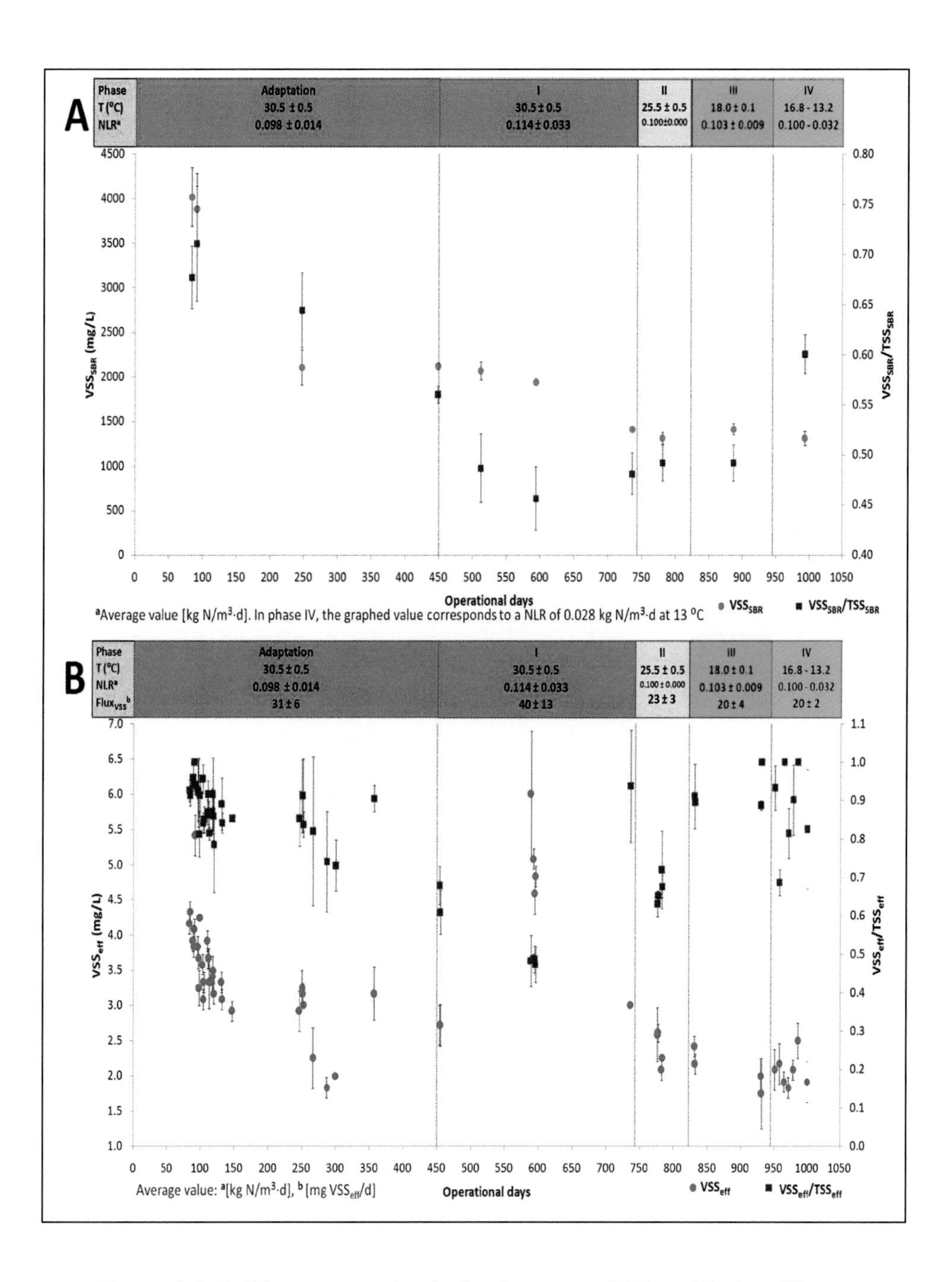

Figure 3.3. Solids concentration in the Anammox SBR and in the effluent.

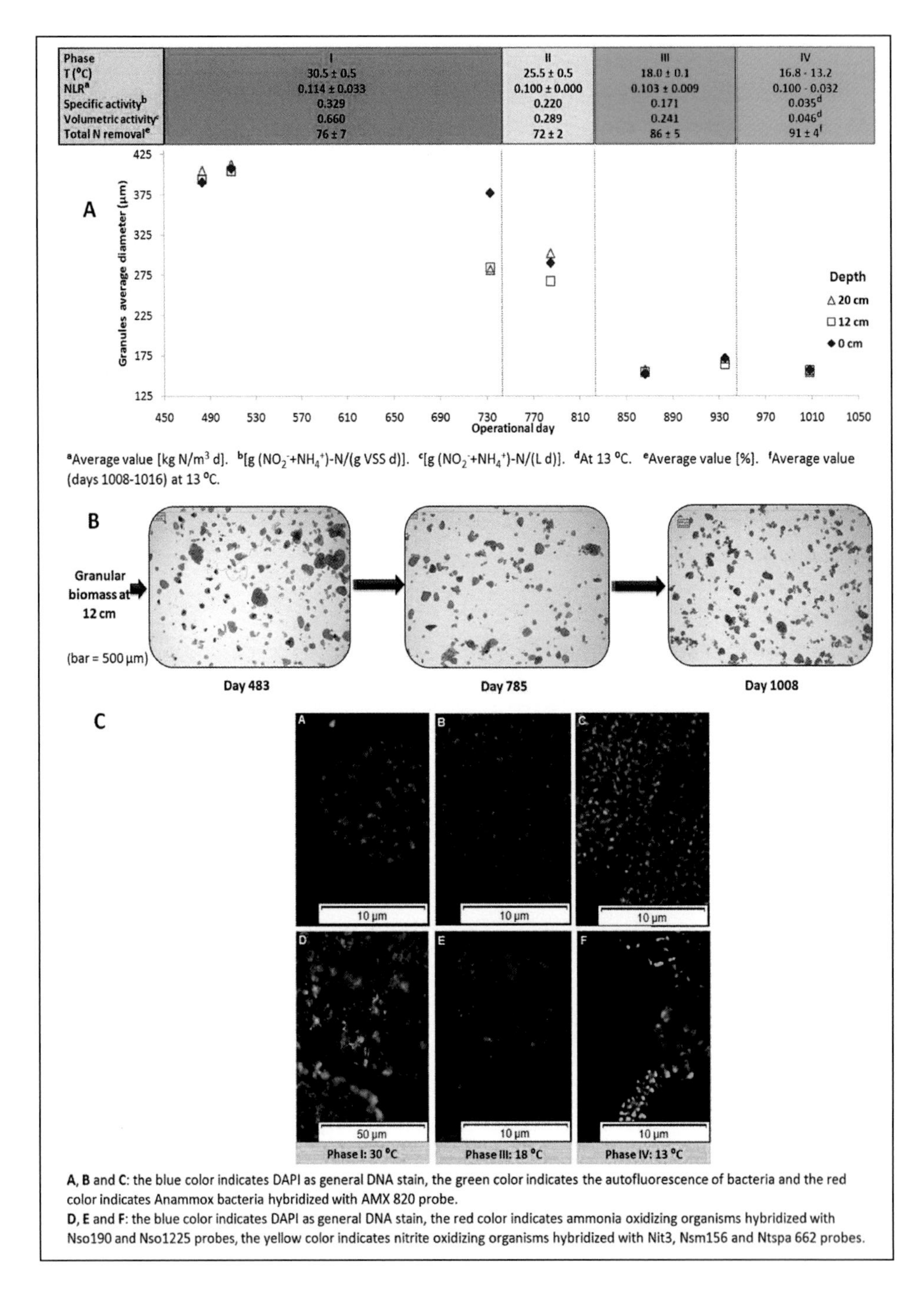

Phase	I	II	III	IV
T (°C)	30.5 ± 0.5	25.5 ± 0.5	18.0 ± 0.1	16.8 - 13.2
NLR[a]	0.114 ± 0.033	0.100 ± 0.000	0.103 ± 0.009	0.100 - 0.032
Specific activity[b]	0.329	0.220	0.171	0.035[d]
Volumetric activity[c]	0.660	0.289	0.241	0.046[d]
Total N removal[e]	76 ± 7	72 ± 2	86 ± 5	91 ± 4[f]

Figure 3.4. Granule size and microbial population variation in the Anammox SBR.

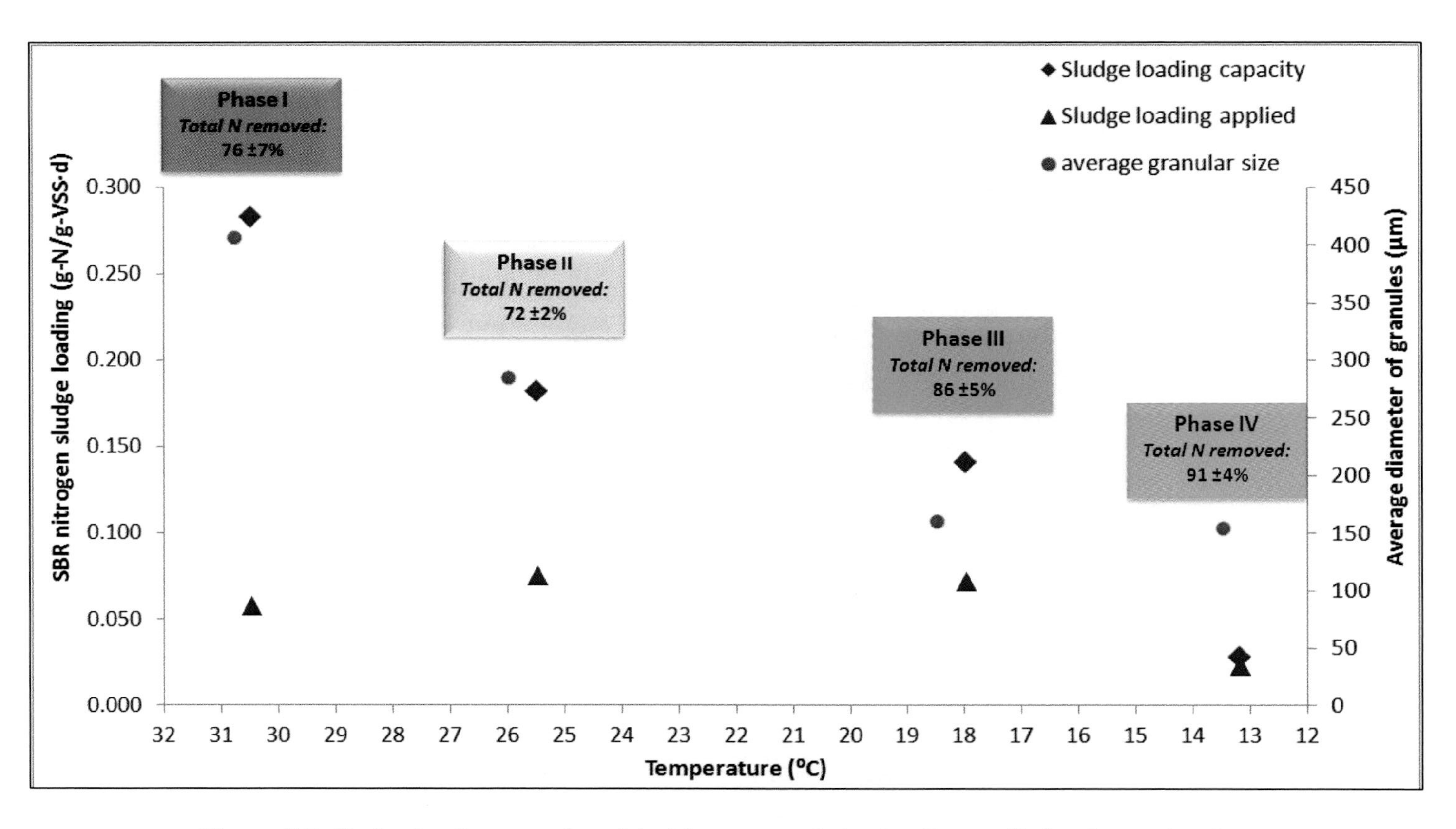

Figure 3.5. Sludge loading capacity of the biomass *vs*. sludge loading applied and granular size.

The total nitrogen removal was 73 ± 2% and the stoichiometric ratios were 1.39 ± 0.09 $g\text{-}NO_2^-\text{-}N/g\text{-}NH_4^+\text{-}N$ and 0.25 ± 0.02 $g\text{-}NO_3^-\text{-}N_{produced}/g\text{-}NH_4^+\text{-}N_{consumed}$. Although the biomass concentration in the reactor remained without variation, the effluent had a significant increase in the TSS_{eff} and VSS_{eff} concentrations (Figure 3.3.B). Before the malfunctioning of the feeding pumps, the values of TSS_{eff}, VSS_{eff} and flux of VSS_{eff} were 4.0 ± 0.3 mg-TSS /L, 2.7 ± 0.3 mg-VSS /L and 26 ± 0.7 mg-VSS/d; but afterwards, these values changed to 11 ± 1.3 mg-TSS/L, 5.1 ± 0.6 mg-VSS/L and 50 ± 6.4 mg-VSS/d.

The effect caused by the NLR increase in the Anammox SBR, on the effluent biomass concentration, could be explained considering the results of Alphenaar (1994), who studied the impact of a sudden drop of 60% in the organic loading rate of a UASB reactor. The UASB reactor was operated at this reduced loading for a period of 56 days. Later, the loading rate was re-established to its original value and the UASB reactor was operated at the restored loading rate for a period of 22 days. As a consequence of these changes in the operational regime, the granular biomass content inside the UASB reactor decreased during the last 22 days of operation. Alphenaar (1994) attributed the observed decrease to a significant disintegration of granules, resulting from the variations in the sludge loading rates.

Similarly, in the Anammox SBR the increase in TSS_{eff}, VSS_{eff} and flux of VSS_{eff} could be explained following the hypothesis that because of the 63% increase in NSLR, the availability of substrate in the granule core increased leading to an increased nitrogen gas production. This increased gas production might have been the driving force that caused a detachment of granular material that promoted the rise of solids concentrations in the effluent. It is remarkable that during the increment of the NSLR, the operational conditions of the SBR allowed to keep the solids concentration in the reactor without variation, given that the original biomass concentration remained constant inside the Anammox SBR, i.e. 1940 ± 40 mg-VSS/L (Figure 3.3.A).

The size of the granules at operational day 733, was similar for the depths at 12 and 20 cm, i.e. an average diameter of 281-284 μm, whereas at the bottom of the reactor (0 cm) the average diameter was 377 μm (Figure 3.4.A). This represents a reduction of around 30% in the size of the granules. Also, at the end of Phase I, the biomass inside the SBR decreased to a concentration of 1413 ± 23 mg-VSS/L. But, the concentration of biomass in the effluent was similar to the value reported at the beginning of Phase

I, i.e. 3.0 ± 0.0 mg-VSS/L. Considering that the mixing conditions remained without changes, the reductions in the granule size and biomass content in the SBR could be ascribed to the low NSLR applied during Phase I (Figure 3.5).

The FISH analysis revealed that Anammox bacteria of genus *Kuenenia* were not present in the Anammox SBR, because the biomass did not hybridize with probe Kst1275. Similarly, the cells did not hybridize with the probe Amx1240, but the hybridization took place with the probe Amx820 (specific probe for the Anammox genus *Kuenenia* and *Brocadia*). Also, an autofluorescence was recorded at an emission wavelength of 520 nm (Figure 3.4.C). On the basis of these findings, the dominant Anammox population in Phase I was presumably *Candidatus Brocadia fulgida* (henceforth referred as *B. fulgida*) which is the same specie identified in the inoculum. Moreover, ammonium oxidizing organisms (AOOs) were present in the SBR and to some extent might have contributed to NH_4^+-N oxidation; on the other hand nitrite oxidizing organisms (NOOs) were not found (Figure 3.4.C).

The presence of AOOs is indicative of dissolved oxygen (DO) availability inside the SBR. The dissolved oxygen concentration in the Anammox SBR was controlled by sparging nitrogen gas inside the liquid phase during 10 minutes at the beginning of each cycle and by a very slow feeding regime, i.e. approximately 8 mL-influent/min. Although the DO in the influent was not removed, the feeding regime allows keeping micro-aerophilic conditions in the Anammox SBR with a maximum DO concentration in the bulk liquid phase of 0.2%. The average total nitrogen removal and the stoichiometric conversion ratios obtained by the Anammox SBR in Phase I were 76 ± 7%, 1.35 ± 0.12 g-NO_2^--N/g-NH_4^+-N and 0.29 ± 0.07 g-NO_3^--$N_{produced}$/g-NH_4^+-$N_{consumed}$.

3.3.3. Phase II

This phase started at the operational day 743 with the dropping of temperature from 30.5 ± 0.5°C to 25.5 ± 0.5°C. The new temperature was kept for the duration of Phase II; meanwhile the other operational parameters remained without alteration with the aim of assessing the performance of the Anammox SBR to a lower temperature. The NSLR applied in Phase II was only 42% of the total NSLR capacity that the biomass had at 25.5 ± 0.5°C (Figure 3.5). Nevertheless, compared to the end of Phase I, there were no substantial changes in the biomass content of the Anammox SBR and in the biomass concentration in the effluent throughout Phase II; that is 1313 ± 70 mg-VSS/L and 2.3 ± 0.0 mg-VSS/L, respectively (Figure 3.3.A-B). Also the total nitrogen removal and the stoichiometry remained close to the values reported in the end of Phase I; that is, a total nitrogen removal of 71% and the stoichiometric ratios of 1.26 g-NO_2^--N/g-NH_4^+-N and 0.31 g-NO_3^--$N_{produced}$/g-NH_4^+-$N_{consumed}$ (Table 3.2). Similarly, the size of the granules at the different depths was in the same range, i.e. 267-302 μm (Figure 3.4.A-B), but the average size of the granules at the bottom of the reactor diminished in comparison with the size at the end of Phase I, i.e. 377 *vs.* 290 μm. The specific activity fell to 0.22 g-(NO_2^- +NH_4^+)-$N_{consumed}$/(g-VSS·d) which corresponds to a decrease of 33% in the specific activity at Phase I.

At day 45 of Phase II (operational day 788) the temperature was re-established to 30°C and the reactor was operated during three days at this temperature in order to evaluate if the dropping of the specific activity was the result of the temperature change. After three days, the specific activity was measured and the value obtained was 0.332 g-(NO_2^--N+NH_4^+)-$N_{consumed}$/(g-VSS·d) which is similar to the activity recorded at 30°C during Phase I. Subsequently, the temperature of the Anammox SBR was set up back to 25.5 ± 0.5°C for the rest of Phase II.

3.3.4. Phase III

Phase III began at day 823 of operation. The operational conditions of the Anammox SBR were similar to Phase II, except for the temperature that was decreased to 18.0 ± 0.1°C. After 43 days of the starting of Phase III (operational day 866) the Anammox granules had an average size of 152-157 μm (Figure 3.4.A); this size represents only 52% of the granule size registered in Phase II on operational day 785. The decrease in the average granule size could be explained mainly by the effect of the NSLR on the granular biomass. Possibly, since the NSLR capacity of the granules remained under-utilized during Phase II, the bacterial starvation in the granules' core could have led to a reduction in the granular size. Chen *et al.* (2013) have demonstrated that the extracellular polymeric substances (EPS) content in the matrix of Anammox granules decreased considerably during the starvation of Anammox biomass. The diminishing of EPS production by Anammox bacteria compromises the integrity of the granules and may also lead to their fragmentation.

On the other hand, it could be stated that the decrease in temperature through the different phases, could promote the availability of substrate to the inner regions of the granules since the conversion rate in their outer parts would be reduced because of the temperature effect. Following this reasoning, the result would be an increase in the granule size. This, however, was not the case for the biomass in the Anammox SBR.

On the contrary, the average diameter of the granules was decreasing from the Adaptation Period and began to stabilize in Phase III. The average diameter of the granules was measured for the second time in Phase III (day 935) and the results were similar to those obtained on day 866, that is 164-171 μm *vs*. 152-157 μm, respectively (Figure 3.4.A). However, Lotti *et al.* (2014a) found that Anammox granular biomass increased their size from 1.5 to 2.1 mm (average ferret diameter) during the operation of a fluidized bed lab-scale reactor, when the temperature was reduced from 20^{0}C to 10^{0}C. Notably, the operation of the fluidized bed lab-scale reactor had a regime of hydraulic mixing that did not damage the granules' integrity and the NSLR applied was enough to cover the NSLR capacity of the biomass in the fluidized bed reactor during the entire experimental period (estimation based on the paper information). Therefore, in the Anammox SBR the decrease in temperature seems not to be a key factor for promoting the granule size or preserving its integrity when the NSLR applied does not satisfy or is not close to the nitrogen loading rate capacity of the biomass. However, in Phase III, the biomass content of the Anammox SBR and of the effluent were 1413 ± 61 mg-VSS/L and 1.8 ± 0.5 mg-VSS/L, correspondingly; these results are similar to those of Phases I and II (Figure 3.3.A-B).

There was a trial for decreasing the temperature in the SBR in Phase III, from 18.0 ± 0.1^{0}C to 10.2^{0}C. The decrease in temperature was performed on day 902 of operation, whereas the other operational conditions remained without variation. The change in temperature had an immediate effect on the microbial activity. In this regard, during nine consecutive cycles (2.25 days) there was no nitrogen removal in the Anammox SBR. This behavior was confirmed by comparing the analytical results from the concentration of NO_2^--N and NH_4^+-N in the effluent and the theoretical accumulative concentration of NO_2^--N and NH_4^+-N in the Anammox SBR with no activity; the results showed that these values were the same.

The concentration of NO_2^--N and NH_4^+-N reached 54 mg-N/L for both compounds at the end of the cycle number nine, i.e. 108 mg-N_{total}/L inside the SBR. The withdrawal of the 2.5 L of effluent was not executed in the cycle number nine; therefore the final volume of 10 L was kept in the SBR. After cycle number ten, the temperature was restored to 18.0 ± 0.1^{0}C and the feeding of substrate was stopped to study the nitrogen removal behavior; biomass was not washed previously. Microbial activity was recorded and in only 18 hours, equivalent to 3 cycles, the concentration of NO_2^--N, NH_4^+-N and NO_3^--N in the SBR were 0, 10 and 10 mg-N/L, respectively. These concentrations correspond to a NO_2^--N removal of 100%, NH_4^+-N removal of 81% and a total nitrogen removal of 81%. The stoichiometry from the nitrogen removal was according to the Anammox metabolic pathway; that is, ratios of 1.22 g-NO_2^--N/g-NH_4^+-N and 0.22 g-NO_3^--$N_{produced}$/g-NH_4^+-$N_{consumed}$. These results demonstrate that the removal capacity and stoichiometry of the biomass were restored to the original capacity registered before the decrease in temperature to 10.2°C. Strikingly, the biomass metabolic activity was recovered in a short period of time, meaning that the sludge had a quick response to the favorable environmental conditions, i.e. temperature. Furthermore, the starved and small Anammox granules were not inhibited by the accumulation of substrates in the SBR.

The average size of the Anammox granules in Phase III was 14-16 times less compared to the size of the starved granules utilized in the research of Carvajal-Arroyo *et al.* (2014), that is 152-171 μm *vs.* 2400 ± 600 μm, respectively. However, the biomass metabolic behavior in the Anammox SBR was similar to the findings of Carvajal-Arroyo *et al.* (2014) because the starved Anammox granules in the SBR were not inhibited by NO_2^--N when they were simultaneously exposed to NH_4^+-N.

Compared to Phase II, in Phase III the specific activity and the volumetric activity diminished by 22% and 17%, respectively. Nevertheless, the total nitrogen removal in Phase III was increased by 10-14% with respect to the previous phases (Table 3.2), giving an average value of 86 ± 5 %. The increase in the total nitrogen removal could be attributed to the fact that the NSLR applied was more close to the NSLR capacity of the system at 18°C (Figure 3.5), i.e. more available substrate per unit of biomass, resulting from the 10% increase in the feeding ratio NO_2^--N/NH_4^+-N that was carried out in Phase III (Table 3.2). Similarly, a lesser mass transfer limitation in the granules, because of the reduction in their size (152-171 μm), could have contributed to the increment in the total nitrogen removal. This hypothesis is in accordance with the findings of Gilbert *et al.* (2013) who reported that the highest Anammox conversion rates and abundance of *B. fulgida* in an Anammox one-step SBR were found in the granular size fraction of 100-315 μm.

FISH analysis demonstrated no shift in the Anammox population of the SBR at 18°C in Phase III, i.e. the dominant Anammox population was presumably *B. fulgida.* In fact, the optimal temperature for *B. fulgida* has been reported in the range of 20-30°C by Hendrickx *et al.* (2014). The AOO were present in the Anammox SBR and the presence of NOO was also identified (Figure 3.4.C). Despite the existence of the NOO, the stoichiometry registered in the last day of Phase III was in accordance with Anammox bacteria metabolism, that is ratios of 1.25 g-NO_2^--N/g-NH_4^+-N and 0.24 g-NO_3^--$N_{produced}$/g-NH_4^+-$N_{consumed}$.

3.3.5. Phase IV

The temperature in the SBR was decreased approximately $1^{o}C$ per week. The NLR and the NSLR were not changed during this phase, except when the biomass activity was severely affected by the reduction in temperature. Phase IV began on day 945 when the temperature was diminished to $16.8 \pm 0.1^{o}C$. On day 952 the temperature was reduced to $16.0 \pm 0.3^{o}C$ and the pH control was stopped because the pH value in the Anammox SBR remained in the range of 7.89 - 8.00 without pH control. The total nitrogen removal efficiencies and the stoichiometric conversion ratios achieved in the last day of operation at $16.8 \pm 0.1^{o}C$ and $16.0 \pm 0.3^{o}C$ were similar to those of the previous phases, i.e. 75%, 1.32 g-NO_2^--N/g-NH_4^+-N, 0.33 g-NO_3^--$N_{produced}$/g-NH_4^+-$N_{consumed}$ and 70%, 1.47 g-NO_2^--N/g-NH_4^+-N, 0.35 g-NO_3^--$N_{produced}$/g-NH_4^+-$N_{consumed}$, correspondingly.

On day 959, the temperature was set to $14.7 \pm 0.3^{o}C$. In the last day at this temperature, the total nitrogen removal dropped to 67% and the stoichiometric conversion ratios were 1.47 g-NO_2^--N/g-NH_4^+-N and 0.36 g-NO_3^--$N_{produced}$/g-NH_4^+-$N_{consumed}$. The Anammox SBR began to become unstable when the temperature was decreased to $13.8 \pm 0.2^{o}C$ while the NLR remained in 0.095 ± 0.013 kg-N/m^3·d. Towards the end of this period, the total nitrogen removal was only 30% and the stoichiometric conversion ratios were 1.20 g-NO_2^--N/g-NH_4^+-N and 0.51 g-NO_3^--$N_{produced}$/g-NH_4^+-$N_{consumed}$. The maximum DO concentration in the SBR remained in 0.2%, but the value of the ratio g-NO_3^--$N_{produced}$/g-NH_4^+-$N_{consumed}$ suggests a higher metabolic activity by the NOOs at $13.8 \pm 0.2^{o}C$ than at $18.0 \pm 0.1^{o}C$ (Phase III). This behavior is supported by the findings of Hellinga *et al.* (1998), which is the growth rate of NOOs is higher than that of AOOs at lower temperatures.

On day 973, the temperature was reduced to $13.2 \pm 0.3^{o}C$. The pH control was reestablished in the Anammox SBR on day 978 and since this day the pH was 7.80 - 8.25 (Table 3.2). The specific activity at $13.2 \pm 0.3^{o}C$ was 0.035 g-(NO_2^- +NH_4^+)-$N_{consumed}$/(g-VSS·d). From this result, the NLR capacity of the Anammox SBR and the NSLR capacity of the biomass were calculated to be 0.038 kg-N/m^3·d and 0.029 g-N/g-VSS·d, respectively. The operational NLR was adjusted to an approximate value of 0.032 kg-N/m^3·d, which corresponded to 83% of the NSLR capacity of the biomass in the Anammox SBR. However, during this stage some difficulties in adjusting the flows in the pumps for the feeding of substrate occurred; this situation caused an unstable operation in the Anammox SBR. As soon as the flows were controlled, the Anammox SBR carried out a total nitrogen removal of 93% on day 1016 with a total nitrogen concentration in the influent (NO_2^--N+NH_4^+-N) of 32 ± 3 mg-N/L as well as

NLR and NSLR set to 0.032 ± 0.004 kg-N/m^3·d and 0.024 g-N/g-VSS·d, respectively (Table 3.2 and Figure 3.5). The corresponding stoichiometric conversion ratios were 1.12 g-NO_2^--N/g-NH_4^+-N and 0.01 g-NO_3^--$N_{produced}$/g-NH_4^+-$N_{consumed}$; the production of NO_3^--N per gram of NH_4^+-N consumed increased to 0.15 g-NO_3^--$N_{produced}$/g-NH_4^+-$N_{consumed}$ on day 1024, which is indicative of Anammox growth (Lotti *et al.*, 2012).

During the last period of Phase IV, i.e. days 1027-1048, a continuing deterioration of the metabolic activity of biomass occurred until there was no nitrogen removal. A white leachate, a product of the reaction between the 0.4 M NaOH solution and the tubing for delivering this solution for pH control in the Anammox SBR presumably caused the inhibition of the metabolic activity. In order to confirm that the leachate had a negative effect in the biomass, the liquid phase in the Anammox SBR was discarded and the biomass was washed two times with 8 L of substrate without (NO_2^--N+NH_4^+)-N. The substrate was sparged previously with nitrogen gas and the final DO concentration was approximately 1%. After the second washing, an additional volume of 8 L of substrate was let overnight inside the reactor and mixed at 180 rpm; nitrogen gas was sparged until the bulk liquid phase reached a DO level of 0.2%. The next day the liquid phase was removed and new substrate was supplied reaching a volume of 7.5 L. From this point the normal operation of the Anammox SBR was restarted. The specific activity was measured after 4 days of operation and the biomass recovered 34% of the specific activity registered before the deterioration of the biomass activity, therefore it was concluded that the leachate was the cause of the metabolic inhibition.

The presence of NOOs, AOOs and Anammox bacteria *B. fulgida* on Phase IV were confirmed by FISH analysis (Figure 3.4.C). Apparently, *B. fulgida* was the dominant Anammox strain throughout the long-term SBR operation at low NSLR and during the decrease in temperature. The ability of *B. fulgida* to acclimate to low temperatures, e.g. down to 10°C, has been reported previously by other authors as Hendrickx *et al.* (2014) and Lotti *et al.* (2014a; 2014b).

During Phase IV, the NSLR capacity of the biomass and NSLR applied were almost the same (Figure 3.5) and the average diameter of the granules was 154-157 μm. This value is similar to the size measured 142 days earlier in Phase III (Figure 3.4.A), suggesting that the stabilization of the size of the Anammox granules can be attributed to the fact that the difference between the NSLR capacity and the applied NSLR was not that high (Figure 3.5). Similarly, the concentration of VSS_{SBR} remained stable since the last days of Phase I until Phase IV (Figure 3.3.A).

3.4. Conclusions

Anammox granular biomass is affected in its integrity and concentration, if the applied NSLR is distinctly lower compared to NSLR capacity of the biomass. Under such conditions, the decrease in temperature seems not to be a key factor for promoting the granule size or preserving its integrity. An adequate load of substrate per unit of biomass and the presence of small granules might increase the total nitrogen removal efficiency. *B. fulgida* was the dominant Anammox strain throughout the long-term operation. Optimization of the NSLR is important for the successful performance of an Anammox granular SBR.

References

Alphenaar A., 1994. Anaerobic Granular Sludge: Characterization, and Factors Affecting its Functioning. PhD dissertation. Wageningen University, pp. 59-72.

Arrojo B., Mosquera-Corral A., Campos J.L., Méndez R., 2006. Effects of mechanical stress on Anammox granules in a sequencing batch reactor (SBR). Journal of Biotechnology, 123, 453-463.

Carvajal-Arroyo J.M., Puyol D., Li G., Swartwout A., Sierra-Álvarez R., Field J.A., 2014. Starved anammox cells are less resistant to NO_2^- inhibition. Water Research, 65, 170-176.

Chen T.T., Zheng P., Shen L.D., 2013. Growth and metabolism characteristics of anaerobic ammonium-oxidizing bacteria aggregates. Environmental Biotechnology, 97, 5575-5583.

Chernicharo, C.A.L., van Lier J.B., Noyola A., Bressani Ribeiro T., 2015. Anaerobic sewage treatment: state of the art, constraints and challenges. Reviews in Environmental Science and Bio/Technology, 14 (4), 649-679.

Dapena-Mora A., Arrojo B., Campos J.L., Mosquera-Corral A., Méndez R., 2004a. Improvement of the settling properties of Anammox sludge in an SBR. J. Chem. Technol. Biotechnol., 79, 1417-1420.

Dapena-Mora A., Campos J.L., Mosquera-Corral A., Jetten M.S.M., Méndez R., 2004b. Stability of the ANAMMOX process in a gas-lift reactor and a SBR. Journal of Biotechnology, 110, 159-170.

Durán U., Val del Río A., Campos J.L., Mosquera-Corral A., Méndez R., 2014. Enhanced ammonia removal at room temperature by pH controlled partial nitrification and subsequent anaerobic ammonium oxidation. Environmental Technology, 35 (4), 383-390.

Gilbert E.M., Müller E., Horn H., Lackner S., 2013. Microbial activity of suspended biomass from a nitritation-anammox SBR in dependence of operational condition and size fraction. Appl. Microbiol. Biotechnol., 97 (19), 8795-8804.

Gilbert E.M., Agrawal S., Karst S.M., Horn H., Nielsen P.H., Lackner S., 2014. Low temperature partial nitritation/anammox in a moving bed biofilm reactor treating low strength wastewater. Environmental Science and Technology, 48, 8784-8792.

Gujer W., 2010. Nitrification and me - A subjective review. Water Research, 44, 1-19.

Hellinga C., Schellen A.A.J.C., Mulder J.W., Van Loosdrecht M.C.M., Heijnen J.J., 1998. The SHARON process: an innovative method for nitrogen removal from ammonium-rich waste water. Water Science and Technology, 37, 135-142.

Hendrickx T.L.G., Kampman C., Zeeman G., Temmink H., Hu Z., Kartal B., Buisman C.J.N., 2014. High specific activity for anammox bacteria enriched from activated sludge at 10°C. Bioresource Technology, 163, 214-221.

ISO 7890/1-1986 (E). Water quality - Determination of nitrate - Part 1: 2,6-Dimethylphenol spectrometric method.

Jetten M.S.M., Strous M., van de Pas-Schoonen K.T., Schalk J., van Dongen U.G.J.M., van De Graaf A.A., Logemann S., Muyzer G., van Loosdrecht M.C.M., Kuenen J.G., 1999. The anaerobic oxidation of ammonium. FEMS Microbiology Reviews, 22, 421-437.

Lackner S., Gilbert E.M., Vlaeminck S.E., Joss A., Horn H., van Loosdrecht M.C.M., 2014. Full-scale partial nitritation/anammox experiences - An application survey. Water Research, 55, 292-303.

Lotti T., van der Star W.R.L., Kleerebezem R., Lubello C., van Loosdrecht M.C.M., 2012. The effect of nitrite inhibition on the anammox process. Water Research, 46, 2559-2569.

Lotti T., Kleerebezem R., van Erp Taalman Kip C., Hendrickx T.L.G., Kruit J., Hoekstra M., van Loosdrecht M.C.M., 2014a. Anammox growth on pretreated municipal wastewater. Environmental Science & Technology, 48 (14), 7874 - 7880.

Lotti T., Kleerebezem R., Hu Z., Kartal B., Jetten M.S.M., van Loosdrecht M.C.M., 2014b. Simultaneous partial nitritation and anammox at low temperature with granular sludge. Water Research, 66, 111 -121.

Lotti T., Kleerebezem R., Hu Z., Kartal B., de Kreuk M.K., van Erp Taalman Kip C., Kruit J., Hendrickx T.L.G., van Loosdrecht M.C.M., 2015. Pilot-scale evaluation of anammox-based mainstream nitrogen removal from municipal wastewater. Environmental Technology, 36 (9), 1167 -1177.

Malamis S., Katsou E., Frison N., Di Fabio S., Noutsopoulos C., Fatone F., 2013. Start-up of the completely autotrophic nitrogen removal process using low activity anammox inoculum to treat low strength UASB effluent. Bioresource Technology, 148, 467-473.

Malovanyy A., Yang J., Trela J., Plaza E., 2015. Combination of upflow anaerobic sludge blanket (UASB) reactor and partial nitritation/anammox moving bed biofilm reactor (MBBR) for municipal wastewater treatment. Bioresource Technology, 180, 144-153.

McKeown R.M., Scully C., Mahony T., Collins G., O'Flaherty V., 2009. Long-term (1243 days), low-temperature (4-15°C), anaerobic biotreatment of acidified wastewaters: Bioprocess performance and physiological characteristics. Water Research, 43, 1611-1620.

Mozumder M.S.I., Picioreanu C., van Loosdrecht M.C.M., Volcke E.I.P., 2014. Effect of heterotrophic growth on autotrophic nitrogen removal in a granular sludge reactor. Environmental Technology, 35 (8), 1027-1037.

NEN, 1983. Photometric determination of ammonia in Dutch system. In: Nederlandse Normen (Dutch Standards), International Organization for Standardization (ed.) NEN 6472, Dutch Institute of Normalization, Delft, the Netherlands.

Ni B.J., Chen Y.P., Liu S.Y., Fang F., Xie W.M., Yu H.Q., 2009. Modeling a Granule-Based Anaerobic Ammonium Oxidizing (ANAMMOX) Process. Biotechnology and Bioengineering, 103 (3), 490-499.

Sánchez Guillén J.A., Yimman Y., Lopez Vazquez C.M., Brdjanovic D., van Lier J.B., 2014. Effects of organic carbon source, chemical oxygen demand/N ratio and temperature on autotrophic nitrogen removal. Water Science and Technology, 69 (10), 2079-2084.

Sánchez Guillén J.A., Cuéllar Guardado P.R., Lopez Vazquez C.M., de Oliveira Cruz L.M., Brdjanovic D., van Lier J.B., 2015a. Anammox cultivation in a closed sponge-bed trickling filter. Bioresource Technology, 186, 252-260.

Sánchez Guillén J.A., Jayawardana L.K.M.C.B., Lopez Vazquez C.M., de Oliveira Cruz L.M., Brdjanovic D., van Lier J.B., 2015b. Autotrophic nitrogen removal over nitrite in a sponge-bed trickling filter. Bioresource Technology, 187, 314-325.

Schmid M., Twachtmann U., Klein M., Strous M., Juretschko S., Jetten M., Metzger J.W., Schleifer K.H., Wagner M., 2000. Molecular evidence for genus level diversity of bacteria capable of catalyzing anaerobic ammonium oxidation. Systematic and Applied Microbiology, 23 (1), 93-106.

Standard Methods for the Examination of Water and Wastewater, 2012, 22 ed. American Public Health Association/American Water Works Association/Water Environment Federation, Washington DC, USA.

Strous M., Heijnen J.J., Kuenen J.G., Jetten M.S.M., 1998 The sequencing batch reactor as a powerful tool for the study of slowly growing anaerobic ammonium- oxidizing microorganisms. Appl. Microbiol., 50 (5), 589-596.

van de Graaf A.A., de Bruijn P., Robertson L.A., Jetten M.S.M., Kuenen J.G., 1996. Autotrophic growth of anaerobic ammonium-oxidizing micro-organisms in a fluidized bed reactor. Microbiology, 142, 2187-2196.

van der Star W.R.L., Abma W.R., Blommers D., Mulder J.W., Tokutomi T., Strous M., Picioreanu C., van Loosdrecht M.C.M., 2007. Startup of reactors for anoxic ammonium oxidation: Experiences from the first full-scale anammox reactor in Rotterdam. Water Research, 41, 4149-4163.

van Loosdrecht M.C.M., 2008. Innovative nitrogen removal. In Biological Wastewater Treatment: Principles, Modelling and Design (M. Henze, M.C.M. van Loosdrecht, G.A. Ekama and D. Brdjanovic, eds.), IWA Publishing, London, UK, pp. 145.

Volcke E.I.P., Picioreanu C., De Baets B., van Loosdrecht M.C.M., 2012. The Granule Size Distribution in an Anammox-Based Granular Sludge Reactor Affects the Conversion - Implications for Modeling. Biotechnology and Bioengineering, 109 (7), 1629-1636.

Appendix 3A

Optimum depth for obtaining a representative sample of MLSS and MLVSS in the Anammox SBR

In Phase 3 of the cycle of operation was performed the test to estimate the optimum depth for obtaining representative samples of MLSS and MLVSS in the reactor. The mixed liquor of the Anammox SBR had a total depth of 26 cm (working volume of 10 L) and the boundaries of the test were set at a total depth of 24 cm. Samples were collected by triplicate at depths of 0, 8, 16 and 24 cm (from the bottom to the top); these samples were used to determine the concentrations of MLSS and MLVSS.

From the average concentrations of MLSS and MLVSS at each depth (Table 3.A.1), the graphs of MLSS *vs*. depth and MLVSS *vs*. depth were plotted (Figures 3.A.1 and 3.A.2, respectively). In both graphs, the best correlation that describes the concentrations of MLSS and MLVSS as a function of the depth is a cubic function. Both cubic functions were used to calculate the theoretical concentrations of MLSS and MLVSS at depth increments of 1 cm (Table 3.A.2).

Similarly, intervals of depths of 1 cm were chosen to discretize the total volume of the mixed liquor. The average concentration and mass content of MLSS and MLVSS were estimated for each discrete volume of 397.6 cm^3 that was defined by the depth interval (Tables 3.A.3 and 3.A.4). To obtain the mean concentrations of MLSS and MLVSS of the total mixed liquor column, the equations 3.A.1 and 3.A.2 were utilized. By using the mean concentrations and figures 3.A.1 and 3.A.2, the optimum depth for sampling the MLSS and MLVSS in the Anammox SBR is obtained; this is 12 cm approximately.

Table 3.A.1. Average values of MLSS and MLVSS at different depths

Running Time (*days*)	Depth (*cm*)	Average MLSS (*mg/L*)	Average MLVSS (*mg/L*)
84	0	7273	5016
	8	6786	4284
	16	5269	3711
	24	4349	3226

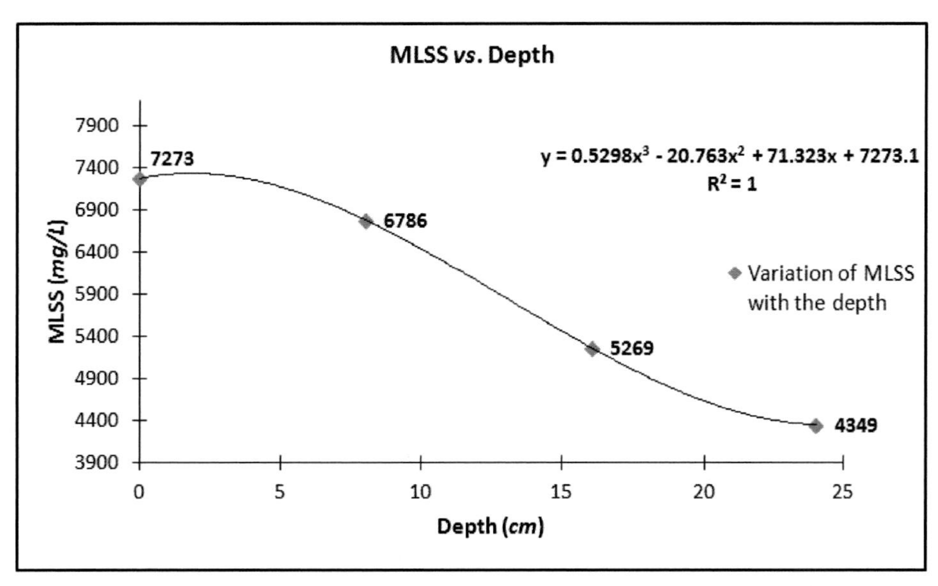

Figure 3.A.1. Average MLSS at different depths in the Anammox SBR.

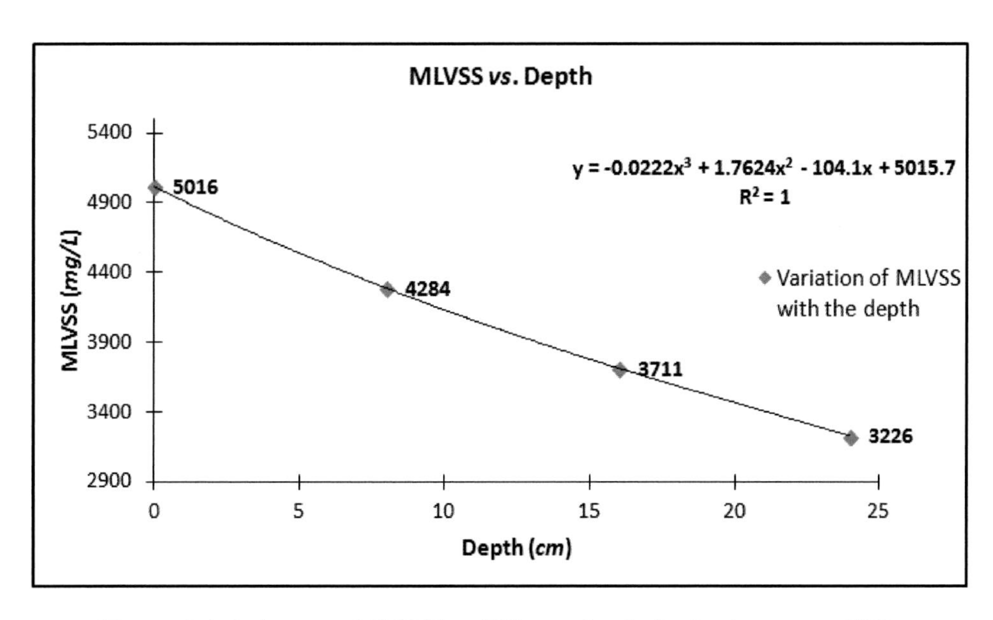

Figure 3.A.2. Average MLVSS at different depths in the Anammox SBR.

Table 3.A.2. Theoretical concentrations of Mixed Liquor Suspended Solids ($MLSS_{th}$) and Mixed Liquor Volatile Suspended Solids ($MLVSS_{th}$) at depth increments of 1 cm.

Depth (cm)	$MLSS_{th}$ (mg/cm^3)	$MLVSS_{th}$ (mg/cm^3)
0	7.273	5.016
1	7.324	4.913
2	7.337	4.814
3	7.315	4.719
4	7.260	4.626
5	7.177	4.536
6	7.068	4.450
7	6.937	4.366
8	6.786	4.284
9	6.619	4.205
10	6.440	4.129
11	6.250	4.054
12	6.055	3.982
13	5.855	3.911
14	5.656	3.843
15	5.459	3.776
16	5.269	3.710
17	5.088	3.646
18	4.919	3.583
19	4.767	3.522
20	4.633	3.461
21	4.521	3.401
22	4.434	3.342
23	4.376	3.284
24	4.349	3.226

Table 3.A.3. Average concentration and mass content of MLSS of each discrete volume.

Depth Interval (cm)	$MLSS_i$ (mg/cm^3)	Δd_i cm	A_i cm^2	$\Delta d_i * A_i * MLSS_i$ mg
0 - 1	7.299	1	397.6	2902
1 -2	7.331	1	397.6	2915
2 -3	7.326	1	397.6	2913
3 -4	7.287	1	397.6	2897
4 -5	7.218	1	397.6	2870
5 -6	7.122	1	397.6	2832
6 -7	7.002	1	397.6	2784
7 -8	6.861	1	397.6	2728
8 -9	6.703	1	397.6	2665
9 -10	6.530	1	397.6	2596
10 -11	6.345	1	397.6	2523
11 -12	6.153	1	397.6	2446
12 -13	5.955	1	397.6	2368
13 -14	5.756	1	397.6	2288
14 -15	5.558	1	397.6	2210
15 -16	5.364	1	397.6	2133
16 -17	5.178	1	397.6	2059
17 -18	5.004	1	397.6	1990
18 -19	4.843	1	397.6	1926
19 -20	4.700	1	397.6	1869
20 -21	4.577	1	397.6	1820
21 -22	4.478	1	397.6	1780
22 -23	4.405	1	397.6	1752
23 -24	4.363	1	397.6	1735

Table 3.A.4. Average concentration and mass content of MLVSS of each discrete volume.

Depth Interval (*cm*)	$MLVSS_i$ (mg/cm^3)	Δd_i *cm*	A_i cm^2	$\Delta d_i * A_i * MLVSS_i$ *mg*
0 - 1	4.965	1	397.6	1974
1 -2	4.864	1	397.6	1934
2 -3	4.767	1	397.6	1895
3 -4	4.672	1	397.6	1858
4 -5	4.581	1	397.6	1822
5 -6	4.493	1	397.6	1787
6 -7	4.408	1	397.6	1753
7 -8	4.325	1	397.6	1720
8 -9	4.245	1	397.6	1688
9 -10	4.167	1	397.6	1657
10 -11	4.092	1	397.6	1627
11 -12	4.018	1	397.6	1598
12 -13	3.947	1	397.6	1569
13 -14	3.877	1	397.6	1542
14 -15	3.809	1	397.6	1515
15 -16	3.743	1	397.6	1488
16 -17	3.678	1	397.6	1463
17 -18	3.615	1	397.6	1437
18 -19	3.553	1	397.6	1413
19 -20	3.491	1	397.6	1388
20 -21	3.431	1	397.6	1364
21 -22	3.372	1	397.6	1341
22 -23	3.313	1	397.6	1317
23 -24	3.255	1	397.6	1294

The mean concentrations of the Mixed Liquor Suspended Solids ($MLSS_m$) and the Mixed Liquor Volatile Suspended Solids ($MLVSS_m$) in the total liquor column were obtained by using the following equations:

Eq. (3.A.1)

$$MLSS_m = \frac{\Sigma \Delta d_i A_i MLSS_i}{A_i \, \Sigma \Delta d_i}$$

Eq. (3.A.2)

$$MLVSS_m = \frac{\Sigma \Delta d_i A_i MLVSS_i}{A_i \, \Sigma \Delta d_i}$$

where,

$MLSS_m$: mean concentration of the Mixed Liquor Suspended Solids based on discrete volume calculation, mg/cm^3

$MLVSS_m$: mean concentration of the Mixed Liquor Suspended Solids based on discrete volume calculation, mg/cm^3

$MLSS_i$: concentration of Mixed Liquor Suspended Solids at *i*th depth interval, mg/cm^3

$MLVSS_i$: concentration of Mixed Liquor Volatile Suspended Solids at *i*th depth interval, mg/cm^3

Δd_i : depth increment about $MLSS_i$ or $MLVSS_i$, cm

A_i : transverse area of the reactor, cm^2

Results

$MLSS_m$ = 5.973 mg/cm^3 = 5973 mg/L ⟶ Depth: 12.4 cm

$MLVSS_m$ = 4.028 mg/cm^3 = 4028 mg/L ⟶ Depth: 11.4 cm

The optimum depth is approximately equal to 12 cm.

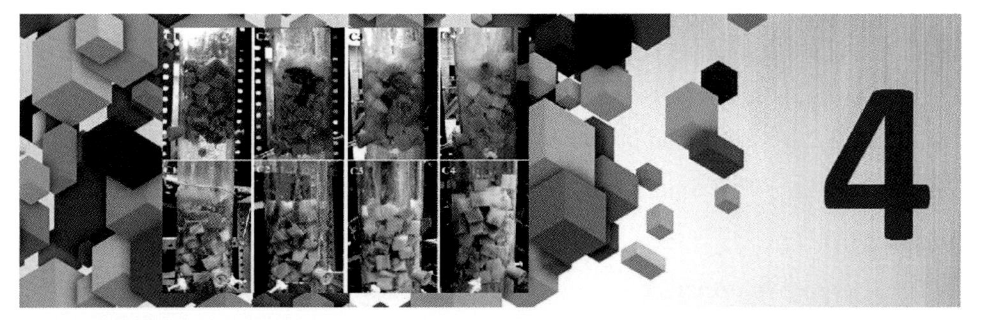

Anammox process in a closed sponge-bed trickling filter

Contents

This chapter is based on:

Sánchez Guillén J.A., Cuéllar Guardado P.R., Lopez Vazquez C.M., de Oliveira Cruz L.M., Brdjanovic D., van Lier J.B., 2015. Anammox cultivation in a closed sponge-bed trickling filter. Bioresource Technology, 186, 252-260.

Abstract

A feasibility study was carried out to assess the cultivation of Anammox bacteria in lab-scale closed sponge-bed trickling filter (CSTF) reactors, namely: CSTF-1 at 20°C and CSTF-2 at 30°C. Stable conditions were reached from day 66 in CSTF-2 and from day 104 in CSTF-1. The early stability of CSTF-2 is attributable to the influence of temperature; nevertheless, by day 405, the nitrogen removal performed by CSTF-1 increased up to similar values of CSTF-2. The maximum total nitrogen removal efficiency was 82% in CSTF-1 and 84% in CSTF-2. After more than 400 days of operation, CSTF-1 and CSTF-2 were capable to attain a total nitrogen removal efficiency of 74 ± 5% and 78 ± 4% with a total nitrogen conversion rate of 1.52 and 1.60 kg-N/m^3_{sponge}·d, respectively. The proposed technology could be a suitable alternative for mainstream nitrogen removal in post-treatment units via Anammox process.

4.1. Introduction

Nitrogen removal via autotrophic nitrite reduction using ammonium as electron donor, better known as the Anammox conversion process, is considered very cost-efficient because of the savings associated to power consumption and low sludge production. However, Anammox bacteria are slow growing organisms that have a low biomass yield. Taking into account this feature, the effective retention of Anammox biomass inside the wastewater treatment facilities is one of the main challenges for the application of the Anammox conversion process, particularly in main-stream treatment lines (Wett *et al.*, 2010). Therefore, in sewage treatment systems the retention of Anammox bacteria is essential to attain: (i) an adequate biodegradation rate through a higher cell density, (ii) a continuous process that can operate at high loading rates and (iii) an efficient and practical separation of the liquid and solid phases in the reactor, leading to simple operation and maintenance.

Several technologies have been applied to cultivate Anammox bacteria in wastewater treatment plants, such as: upflow granular sludge bed reactors (van der Star *et al.*, 2007), sequencing batch reactors (SBR) (Joss *et al.*, 2009), sequencing batch reactor with a cyclone (Wett *et al.*, 2010), airlift reactors (Abma *et al.*, 2010) and upflow anaerobic sludge bed reactors (Ma *et al.*, 2013). Other alternative systems have been proposed the apply Anammox biomass entrapped in a polyethylene glycol (PEG) gel carrier (Isaka *et al.*, 2008) or use membrane bioreactors for the cultivation of Anammox microorganisms as free cells (Lotti *et al.*, 2014).

Also biofilm systems have been applied, using different configurations, for instance, fixed film bioreactors with plastic carrier material and rotating biological contactors (Pynaert *et al.*, 2004), reactors based on hydrophilic net-type acryl fiber biomass carrier (Furukawa *et al.*, 2006), membrane-aerated biofilm reactors (Pellicer-Nacher *et al.*, 2010), moving-bed biofilm reactors (Lackner and Horn, 2013), and sequencing batch biofilm reactors (Yu *et al.*, 2012).

With regard to Anammox biofilm reactors, the use of sponge media as carrier material appears to be a promising alternative that has been explored by some researchers. For instance, Chuang *et al.* (2008) have reported the immobilization and growing of Anammox bacteria in a closed down-flow hanging sponge (DHS) reactor for autotrophic nitrogen removal. On the other hand, different configurations of sponge-based Anammox reactors where the sponges are permanently immersed in the liquid phase have been tested, e.g. the up-flow Anammox column reactor with polyethylene sponge strips mounted in a vertical shaft (containing a significant amount of suspended Anammox granules accumulated in the bottom) (Zhang *et al.*, 2010) and an anaerobic upflow fixed bed reactor (UFBR) developed by Monballiu *et al.* (2013). According to Zhang *et al.* (2010) the Anammox reactor packed with sponges achieved a higher nitrogen conversion rate when compared to other reactors with packing materials like non-woven material and acrylic fiber. The high conversion efficiency was attributed to the high biomass retention capacity of the sponge media. Despite of these potential advantages, only the previous examples of Anammox biofilm reactors with sponges as support material have been found in the literature.

Recent studies have reported the presence of Anammox bacteria in aerobic/anoxic trickling filters designed and operated for nitrification/denitrification purposes. Lydmark *et al.* (2006) observed certain Anammox cells in the biofilm of a municipal full-scale aerated nitrifying trickling filter packed with plastic cross-flow media as biomass carrier. Similarly, Almeida *et al.* (2013) also detected the presence of Anammox microorganisms (abundance not reported) in a pilot-scale municipal trickling filter with sponge packing material (TF-Rotosponge) as support media.

However, the abundance of Anammox bacteria in those reactors was relatively low since they were not aimed to achieve nitrogen removal via Anammox. Recently, Wilsenach *et al.* (2014) identified the presence of Anammox *Candidatus Brocadia anammoxidans* and *Candidatus Brocadia Fulgida* on the biofilm of trickling filters with stones as support media, in the Daspoort municipal wastewater treatment plant, South Africa. According to the authors, the contribution of the Anammox bacteria seems to be important for the nitrogen removal in these trickling filters.

Considering the potential Anammox biomass retention capacity of the sponge media and the occurrence of Anammox bacteria in trickling filter systems, a trickling filter designed and operated with a sponge carrier media can be a promising and efficient technological option for nitrogen removal from sewage by Anammox since it: (i) provides a suitable surface area for biomass growth, (ii) has a high biomass retention capacity and substrate permeability, (iii) is a relatively low energy consumption process and (iv) is a simple technology with low operational and maintenance requirements. Thus, the present research aimed to assess the feasibility to cultivate Anammox bacteria in a closed sponge bed trickling filter, which could eventually contribute to the development of a suitable post-treatment technology for nitrogen removal after anaerobic sewage treatment (Chernicharo *et al.*, 2015).

4.2. Materials and methods

4.2.1. Configuration of the reactors

Two lab-scale closed sponge-bed trickling filters (CSTF) were constructed for this study, namely reactors CSTF-1 and CSTF-2. They were operated simulating moderate and tropical weather conditions at 20 and 30°C, respectively. Each reactor was 100 cm high and had an inner diameter of 7.14 cm. The total volume of each reactor was 4.3 L.

The CSTF reactors were composed of four compartments filled randomly with biomass support media consisting of 95 polyurethane sponge cubes (size: 1.5x1.5x1.5cm) per compartment (Figure 4.1), which is equal to 380 sponge cubes per reactor. Taking into account the total number of sponge cubes per reactor and the volume of each sponge cube (3.375 cm^3), the volume occupied by the sponges inside the CSTF reactors is 1.28 L, i.e. the working volume (based on the occupancy of the sponge volume) is equivalent to 30% of the reactors' volume. The polyurethane sponge media used in this study is BVB Sublime (second generation) (BVB Substrates; De Lier; the Netherlands). The sponge has a void ratio of 0.98 and a density of 28 kg/m^3.

The specific surface area of the support media was estimated as the ratio between the total surface areas of all sponge cubes per unit volume of all the sponge cubes inside the CSTF reactors without considering the microstructure of the sponges (pore size and structure). Therefore, the specific surface area is 400 m^2/m^3. According to the information provided by the manufacturer, the polyurethane sponge has both a vertical and horizontal structure, which results in a good and stable distribution of the influent and good drainage. Also, it has a high dimensional stability (no shape deformation) and is innocuous against chemicals.

The reactors were operated under anoxic conditions; therefore, no oxygen was supplied. In addition, to eject and avoid the presence of oxygen in the demineralized water fed to the reactors, from day 43 onwards, nitrogen gas was sparged into the demineralized water feeding tank and a water lock (containing a solution of sodium sulfite and cobalt chloride) was connected to the head space.

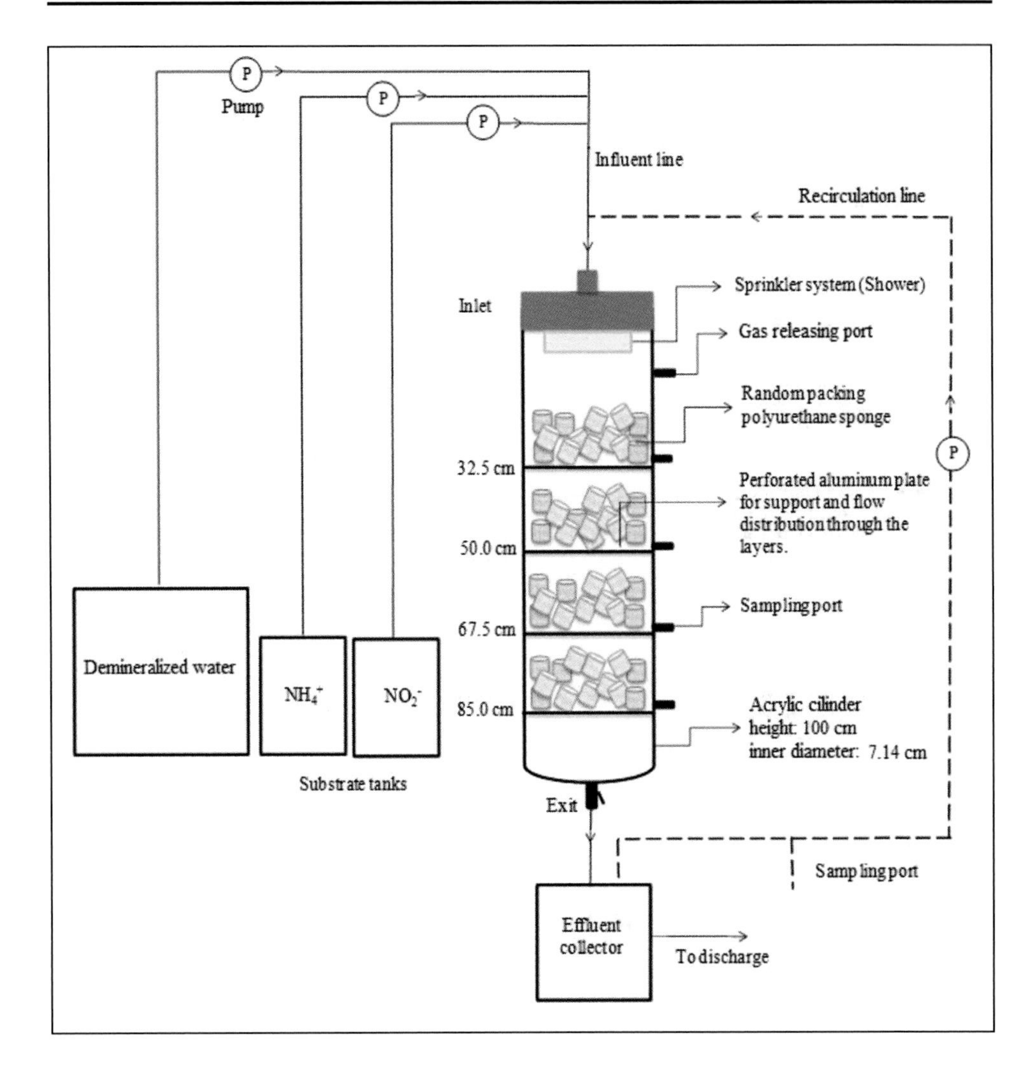

Figure 4.1. Closed sponge-bed trickling filter (CSTF) scheme.

4.2.2. Inoculation and operational conditions

Anammox granular sludge from the lab-scale SBR described in chapter 3 was used as seed. At the moment of sampling, day 393, the SBR was operated at 30.5 ± 0.5°C and pH of 7.89 ± 0.01, with a mixed liquor suspended solids (MLSS) concentration of 3290 mg/L. According to FISH analysis, the dominant Anammox microorganisms in the SBR were *Candidatus Brocadia fulgida* with a specific activity of 0.34 g-(NO_2^- +NH_4^+)-$N_{consumed}$/(g-VSS·d) and an average granule diameter of 0.394 ± 0.244 mm. In addition, activated sludge from Harnaschpolder wastewater treatment plant (Den Hoorn, The Netherlands) with an average concentration of MLSS of 3500 mg/L was also added inside the CSTF reactors to allow the potential growth of other (Anammox) species (if they could benefit the applied environmental and operating conditions). Each reactor's compartment was filled randomly with 45 polyurethane sponge cubes previously saturated with synthetic substrate. Later on, 125 mL of Anammox granular sludge and 12.5 mL of activated sludge were distributed on the sponges. The remaining 40 sponge cubes, also saturated with synthetic substrate, were placed over the microorganisms in each compartment. The reactors were sealed to make them air tight, covered with a dark plastic to avoid light penetration, and nitrogen gas was sparged for 2 h to eject the oxygen present. The reactors were operated under two phases, namely start-up and phase I, as shown in Table 4.1.

4.2.3. Mineral medium

The mineral medium was modified from that one used by van de Graaf *et al.* (1996) and divided in two containers (Figure 4.1): i.e. ammonium rich and nitrite rich feeds. The composition of these substrates per 1 liter of demineralized water was (i) ammonium feed: 2.9828 g NH_4Cl; 0.77 g $MgSO_4{\cdot}7H_2O$; 0.3906 g KH_2PO_4; 4.6875 g $CaCl_2{\cdot}2H_2O$; (ii) nitrite feed: 3.8504 g $NaNO_2$; 0.1786 g $FeSO_4{\cdot}7H_2O$; 19.531 g $KHCO_3$; 0.1786 g NaEDTA and 1.25 mL of trace element solution. The trace element solution contained per liter: 15 g Mg EDTA; 0.43 g $ZnSO_4{\cdot}7H_2O$; 0.24 g $CoCl_2{\cdot}6H_2O$; 0.99 g $MnCl_2{\cdot}4H_2O$; 0.25 g $CuSO_4{\cdot}5H_2O$; 0.22 g $Na_2MoO_4{\cdot}2H_2O$; 0.19 g $NiCl_2{\cdot}6H_2O$; 0.1076 g Na_2SeO_4; 0.014 g H_3BO_3; 0.05 g $NaWO_4{\cdot}2H_2O$.

4.2.4. Tracer test

Tracer tests to estimate the hydraulic retention time (HRT; based on sponge volume) were carried out using a lithium chloride solution as tracer (8.5 mg-Li/L). The method described by Metcalf and Eddy (2003) was applied and the tracer was added continuously to the reactor through the influent demineralized water line and mixed with the recirculation flow and synthetic substrates. During the tracer test, the effluent collection and recirculation flow were separated by using a provisional effluent collector tank, while keeping the original effluent collector for the recirculation flow (Figure 4.2). Samples were collected every 15 min at the effluent of the reactor and acidified to pH 1-2 pH with HNO_3. The lithium concentrations were measured by an atomic absorption spectrophotometer AANALYST 200; Perkin Elmer, USA. The lithium concentration as a function of time was used to plot the tracer response curve or C curve.

Table 4.1. Environmental and operational parameters of the CSTF reactors.

Parameter	Unit	CSTF-1		CSTF-2	
		Start-up	**Phase I**	**Start-up**	**Phase I**
Influent flow rate	L/d	14.4	27.0	13.8	27.0
Recirculation flow rate	L/d	-----	27.4	-----	27.6
Influent NH_4^+-N concentration	mg-N/L	50	50 ± 6	50	49 ± 4
Influent NO_2^--N concentration	mg-N/L	50	52 ± 4	50	50 ± 5
NLR*	kg-N/m³·d	1.13	2.15 ± 0.21	1.08	2.09 ± 0.19
Nominal Hydraulic Retention Time (HRT)*	h	2.13	1.14	2.23	1.14
Hydraulic loading rate (HLR)	m³/m²·d	3.6	13.7	3.4	13.7
Temperature	°C	20	20	30	30
Duration	days	16	405	16	404

*Based on sponge volume, i.e. $kg\text{-}N_{influent}/m^3_{sponge}\cdot d$

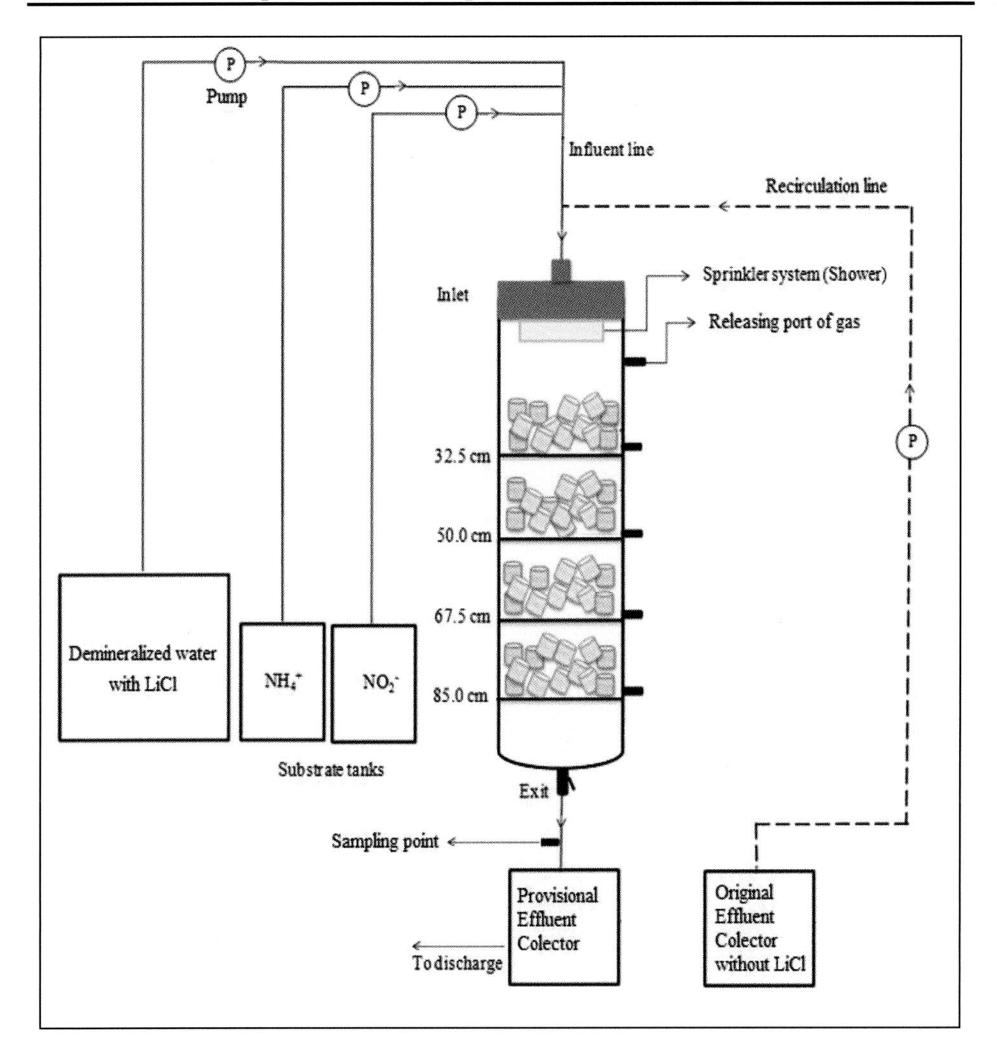

Figure 4.2. Scheme of the closed sponge-bed trickling filter (CSTF) during the tracer tests.

4.2.5. Analytical methods

The total suspended solids (TSS), volatile suspended solids (VSS), soluble chemical oxygen demand (COD) and nitrite (NO_2^--N) concentrations were determined according to Standard Methods for the Examination of Water and Wastewater (2012.). Ammonia (NH_4^+-N) was measured spectrophotometrically following the standard NEN 6472 (NEN, 1983). For nitrate (NO_3^--N) measurements the standard ISO 7890-1:1986 (ISO 7890/1, 1986) was applied. pH was measured with a portable pH meter (Model ProfiLine 3310. WTW, Germany) and the dissolved oxygen (DO) concentrations were recorded using a Hach portable meter HQ30d equipped with the LDO101optical dissolved oxygen probe (Hach Company, USA). The average equivalent granule diameter of the Anammox inoculum was measured with a microscope Leica Microsystems M205 FA (magnification 13.0, calibration factor 4.45 and software version Qwin V3.5.1.; Leica Microsystems Ltd, Netherlands). Anammox bacteria were identified by Fluorescence *in Situ* Hybridization (FISH) technique. Combined biomass samples were scratched out from different sponges of each section and fixed in 4% (w/v) paraformaldehyde solution. Hybridization with fluorescent probes was performed as described by Schmid *et al.* (2000). Epifluorescence was used for cells identification and DAPI (4', 6'-diamidino-2-phemylindol) as general DNA stain. Oligonucleotide probes were labeled with either fluorochromes Cy3 or Cy5 (Biomers.net, Germany). Images were acquainted by an epifluorescence microscope BX51 with a camera XM10 (Olympus, Japan). The probes used in this study were Amx820 (Schmid *et al.*, 2001; Schmid *et al.*, 2003), Amx1240 and Kst1275 (Schmid *et al.*, 2005).

4.3. Results and discussion

4.3.1. Immobilization of the Anammox organisms

After inoculation, Anammox sludge remained attached to the sponge support media and expanded inside and outside the sponge carriers throughout the experiment (Figure 4.3). Overall, the biomass growth was more pronounced in the upper compartments than in the last one, showing a higher nitrogen removal activity in the first compartments. Biomass expansion was faster in CSTF-2 (operated at 30°C) than in CSTF-1 (at 20°C), resulting in biofilm growth on the walls of the compartments number 1 and 2. This likely occurred because of the 30°C temperature applied to CSTF-2, which is closer to the optimum for Anammox microorganisms, i.e. 35°C (Dosta *et al.*, 2008).

With regard to the FISH analysis, the cells were not hybridized with probe Kst1275 (*Candidatus Kuenenia stuttgartiensis*), therefore the Anammox genus *Kuenenia* was not present. No hybridization took place with the probe Amx1240 (*Candidatus Brocadia anammoxidans*), but the hybridization using the probe Amx820 (specific for *Kuenenia* and *Brocadia* anammox bacteria) was carried out and autofluorescence at an emission wavelength of 520 nm was recorded. For these reasons, like in the original inoculum, *Candidatus Brocadia fulgida* was presumably the dominant Anammox population in both reactors (Figure 4.4). The tracer tests indicated that the CSTF-1 and CSTF-2 were operated at an HRT of 1.05 and 1.20 h, respectively. These values are significantly close to the nominal HRT based on sponge volume, i.e. 1.14 h, only 8% lower for CSTF-1 and 5% higher for CSTF-2. Throughout the execution of the experiments, the average pH of the effluent in CSTF-1 increased from 7.2 to 7.6, whereas it increased from 7.4 to 7.9 in CSTF-2. During the entire period of the research, the total suspended solids (TSS) concentrations in the effluent were relatively low in both reactors (13 to 50 mg/L).

4.3.2. Nitrogen profiles

Both CSTF reactors were continuously operated for more than 400 days. The removal of ammonia and nitrite by Anammox bacteria was unstable before day 49, presumably because of low Anammox activity caused by oxygen intrusion through the influent. To avoid and reduce the oxygen intrusion, oxygen scavenger measures were implemented as is described in section 2.1. After the implementation of these measures the dissolved oxygen concentration in the influent dropped from 7.0 mg/L to 1.1 ± 0.2 mg/L and a higher Anammox activity was observed in both reactors. These results suggest that the Anammox activity was inhibited by the presence of oxygen and the metabolic pathway was dominated by the Ammonia Oxidizing Organisms (AOO) and the Nitrite Oxidizing Organisms (NOO) during the first days of operation. This hypothesis is supported by the NO_3^--N concentration registered in the effluents of CSTF-1 and CSTF-2 before day 49, i.e. 46 ± 3 mg-NO_3^--N/L and 43 ± 23 mg-NO_3^--N/L, respectively. Throughout the operation of the CSTF reactors, the nitrite removal activity was higher than the ammonium removal activity, which was expected according to the Anammox stoichiometry (Lotti et al., 2014). The nitrogen compound profiles measured along the CSTF reactors are shown in Figures 4.5 and 4.6.

After day 104, CSTF-1 became stable. From this day up to the end of the research, the average concentration of NH_4^+-N and NO_2^--N in the effluent were 9 ± 3 and 4 ± 4 mg/L, respectively. With regard to NO_3^--N, the average concentration was 31 ± 15 mg/L between day 45 and 66. But from day 104, the average concentration of NO_3^--N decreased significantly to 13 ± 6 mg/L. The performance of CSTF-2 became stable since day 66. From day 104 onwards, the NH_4^+-N and NO_2^--N average concentrations in the effluent of CSTF-2 were 3 ± 3 and 3 ± 1 mg/L, respectively. In terms of the NO_3^--N produced, the average concentration between day 45 and 66 was 11 ± 6 mg /L. A similar NO_3^--N production remained for the rest of the study with an average value of 15 ± 3 mg/L.

Figure 4.3. Anammox microorganisms attached to the support media of CSTF-1 and CSTF-2 on day 79. C1: compartment 1, C2: compartment 2, C3: compartment 3, C4: compartment 4.

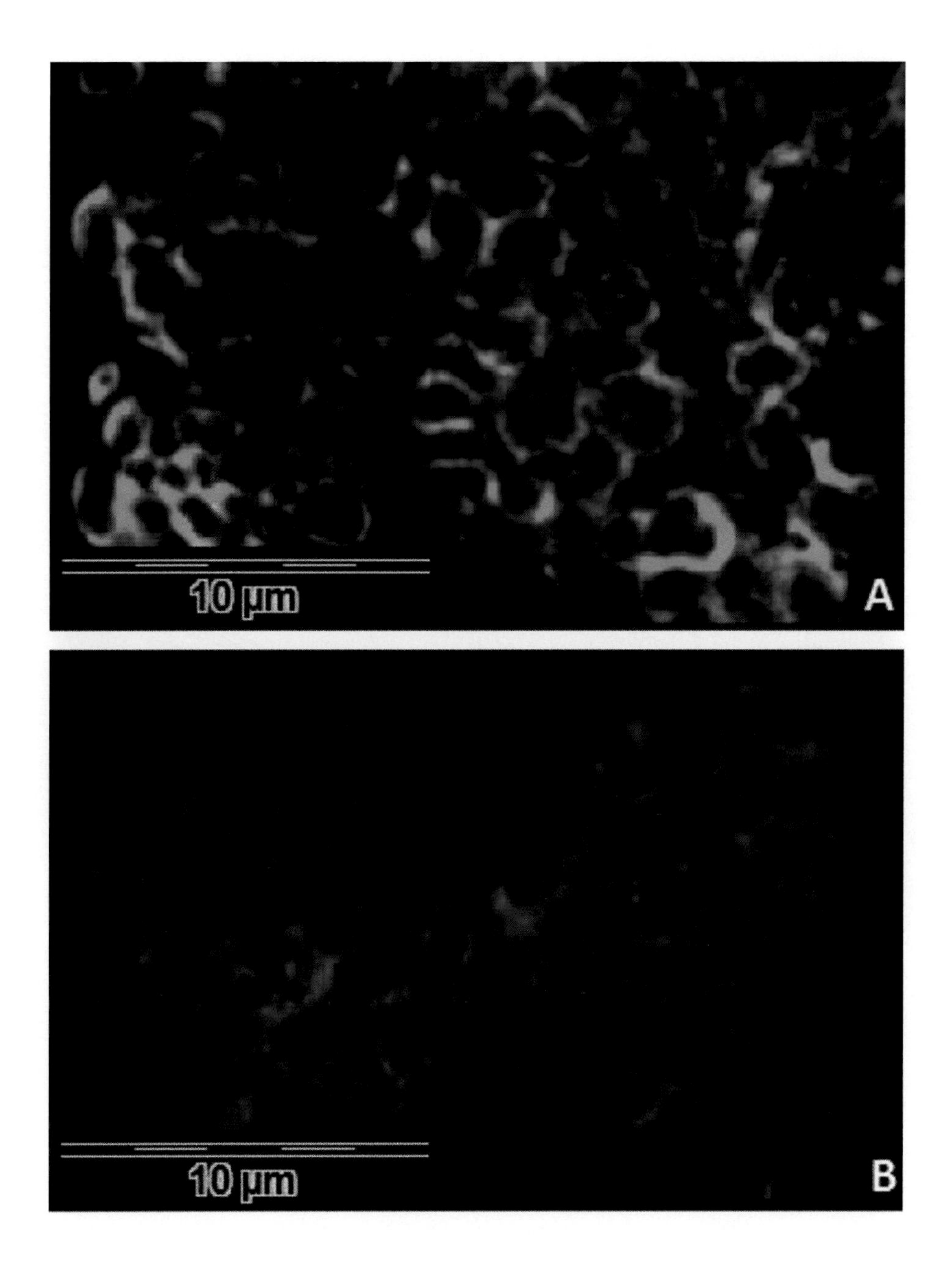

Figure 4.4. Identification of microbial population by FISH analysis. **A**: CSTF-2, **B**: CSTF-1. Red color indicates Anammox bacteria hybridized with Amx820 probe. Green color indicates the autofluorescence of bacteria.

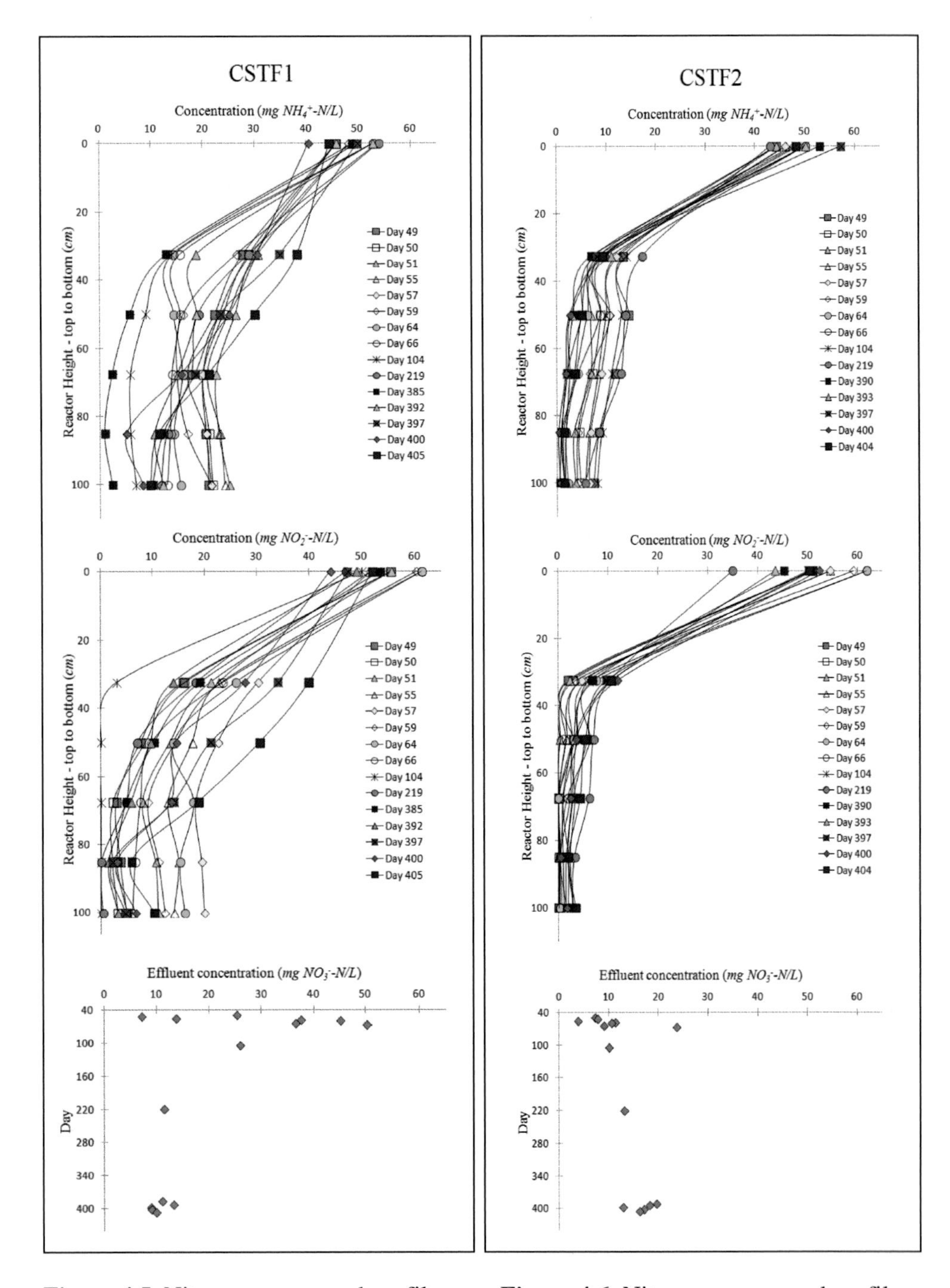

Figure 4.5. Nitrogen compound profiles measured along CSTF-1

Figure 4.6. Nitrogen compound profiles measured along CSTF-2

4.3.3. Nitrogen removal efficiencies and stoichiometric ratios

The evolution of the removal of NH_4^+-N, NO_2^--N and Total-N in CSTF-1 and CSTF-2 is shown in Figure 4.7. In accordance to van Hulle *et al.* (2011), a slow feeding regime is required to favor the Anammox activity and to avoid nitrite peaks that can inhibit Anammox bacteria. In the present study, the CSTF reactors were inoculated with biomass from the lab-scale SBR (chapter 3) working with an NLR of 0.098 ± 0.014 kg-N/m^3_{sponge}·d (NO_2^--N/ NH_4^+-N feeding ratio equal to 0.92 ± 0.10).

Immediately after the inoculation, the startup period began with an NLR of 1.13 kg-N/m^3_{sponge}·d (CSTF-1) and 1.08 kg-N/m^3_{sponge}·d (CSTF-2) for 16 days. Hereafter, the NLR was increased up to 1.90-2.36 kg-N/m^3_{sponge}·d in both reactors. Therefore, besides the time required for the Anammox bacteria to get acclimatized to the CSTF conditions, the sudden change in the feeding regime also may have affected the activity of the Anammox microorganisms at the start-up phase of the experiments. The lower nitrogen removal capacity at the beginning of phase I (Figure 4.7) and the distortion of the stoichiometry (Figure 4.8) could be a reflection of those effects.

The influent DO concentration was another factor that also could have affected the Anammox performance during the first weeks of the tests. For instance, in CSTF-2, a decrease in the total nitrogen removal efficiency was observed possibly due to the oxidation of ammonium (or nitrite) to nitrate. However, after day 43, when the DO control measures in the influent were implemented (i.e. nitrogen sparging of the demineralized water and the water lock), an increased Anammox activity was observed (Table 4.2).

In addition, between day 45 and 66 of operation, the average nitrogen conversion ratios for CSTF-2 were 1.35 ± 0.12 g-NO_2^--N/g-NH_4^+-N and 0.26 ± 0.13 g-NO_3^--$N_{produced}$/g-NH_4^+-$N_{consumed}$. These nitrogen conversion ratios resembled the stoichiometric ratios reported for enriched Anammox cultures, e.g. 1.11-1.54 g-NO_2^--N/g-NH_4^+-N, and 0.16 - 0.26 g-NO_3^--$N_{produced}$/g-NH_4^+-$N_{consumed}$ (Strous *et al.*, 1998; Hendrickx *et al.*, 2012; Lotti *et al.*, 2014). From day 104, the CSTF-1 and CSTF-2 had average ratios of 1.13 ± 0.06 and 0.94 ±0.12 g-NO_2^--N /g-NH_4^+-N, whereas the g-NO_3^--$N_{produced}$/g-NH_4^+-$N_{consumed}$ ratios were 0.28 ± 0.13 and 0.31 ± 0.09, correspondingly. These results suggest that the Anammox metabolism was the dominant nitrogen removal pathway in CSTF-2 from day 45 onwards and in CSTF-1 from day 104 onwards.

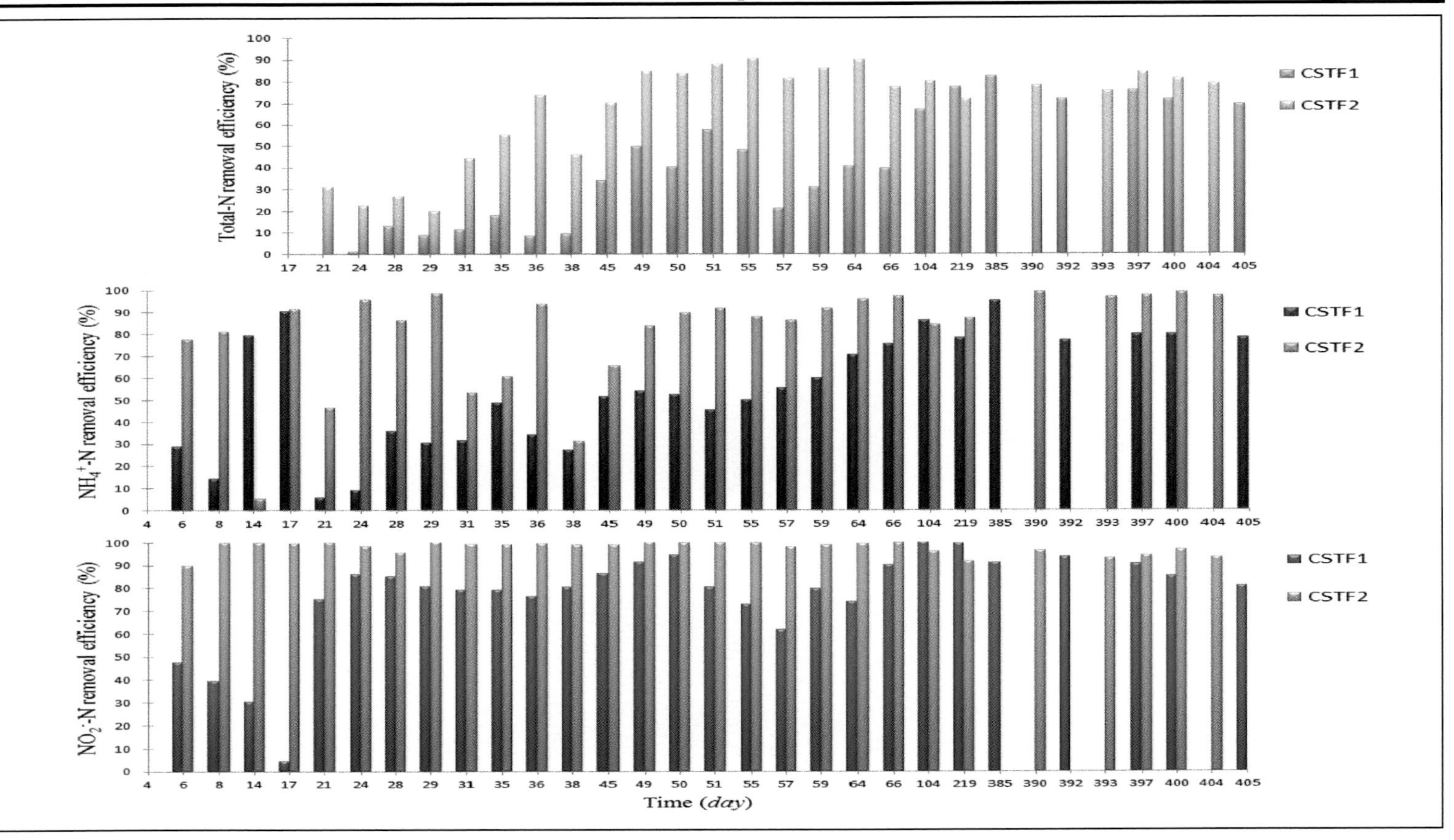

Figure 4.7. Nitrogen removal efficiencies observed in the CSTF reactors.

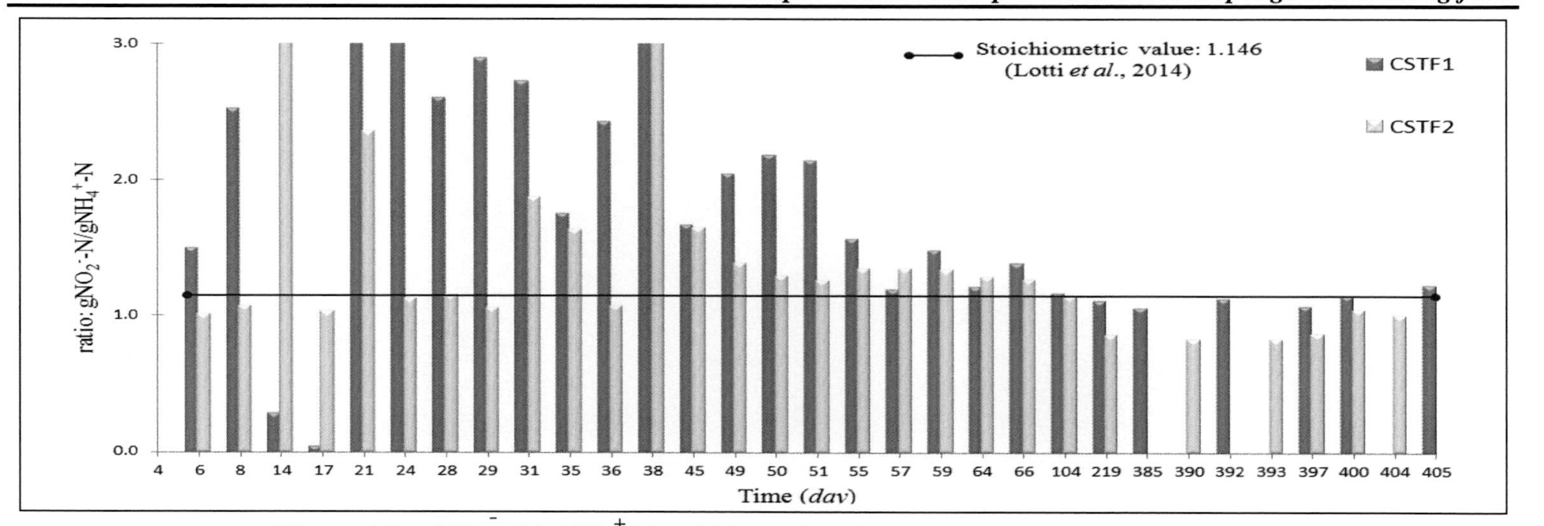

Figure 4.8. g-NO_2^--N/g-NH_4^+-N stoichiometric ratios observed in the CSTF reactors.

Table 4.2. Average nitrogen removal efficiencies of the CSTF reactors.

Period	Average nitrogen removal efficiency (%)					
	Total-N		**NH_4^+-N**		**NO_2^--N**	
	CSTF1	CSTF2	CSTF1	CSTF2	CSTF1	CSTF2
Between day 45 and 104	40 ±12	83 ±7	57 ±10	88 ±9	81 ±11	99 ±1
From day 104 up to the end of the study	74 ±5	78 ±4	82 ±7	92 ±7	91 ±7	94 ±2

4.3.4. Overall operating conditions

Overall, CSTF-2 reached its stability and highest performance faster than CSTF-1. Both CSTF reactors were inoculated with Anammox biomass from the lab-scale SBR operated at 30°C (Chapter 3). Therefore, in CSTF-2 the biomass did not experience any temperature change, whereas the Anammox bacteria in CSTF-1 had to adjust and acclimatize to the lower temperature of 20°C.

Nevertheless and remarkably, CSTF-1 reached a total nitrogen removal similar to CSTF-2 from day 104 (i.e. 74 ± 5% for CSTF-1 and 78 ± 4% for CSTF-2) (Table 4.2). Both CSTF's achieved very high the biomass retention, i.e. low TSS concentration in the effluent (section 4.3.1). In addition, a high nitrogen removal capacity was obtained in the CSTF reactors despite of the change in the NLR, from the Start-up to Phase I, when the concentrations of ammonium and nitrite in the influent were kept constant, while the influent flow was increased, simulating the flow rate changes that occur in full-scale wastewater treatment plants.

In Table 4.3, some operational aspects and the performance of the CSTF reactors are compared to other Anammox biofilm systems. The high Anammox biomass retention capacity achieved by the CSTF reactors provides important advantages. For instance, in comparison to Anammox biofilm reactors (Table 4.3), the short hydraulic retention time (HRT) attained in the CSTF (1.05 -1.20 h) with a volumetric nitrogen conversion rate (per sponge volume) similar to a full-scale system (1.6 kg-N/m^3d; Veuillet *et al.*, 2014) could allow the use of a compact reactor of a small footprint for nitrogen removal. This feature could be especially convenient in regions of low land availability. Also, a compact reactor reduces the investment costs and facilitates the operation and maintenance of the systems, which are suitable characteristics for (decentralized) wastewater treatment plants, particularly in developing countries.

Another potential benefit from the Anammox biomass cultivation in the CSTF reactors is the total nitrogen removal capacity attained under the applied NLR and temperatures. Our results showed that CSTF reactors can reach similar nitrogen removal rates but at a lower temperature, compared to lab-scale Closed DHS (Chuang *et al.*, 2008), lab-scale Sequencing Batch Biofilm reactors (SBBR) (Yu *et al.*, 2012) and Integrated Fixed-film Activated Sludge (IFAS) full-scale systems (Veuillet *et al.*, 2014). It must be noted that the IFAS ANITA™ Mox, SBBR and Closed DHS reactors were operated at 30 ± 3, 35 ± 1 and 35°C, respectively.

Table 4.3. Performance of the CSTF reactors vs. a selection of Anammox biofilm reactors

Type of Reactor	Biomass Carrier	Inoculum	Temp. °C	HRT h	NLR kg-N/m^3 d	N Conversion Rate kg-N/m^3 d	Total N Removal %	Reference
[a]CSTF-1	Polyurethane sponge	Anammox granules and activated sludge	20	1.05	[d]2.15 ± 0.21	[d]1.52	74 ± 5	This study
[a]CSTF-2	Polyurethane sponge	Anammox granules and activated sludge	30	1.20	[d]2.09 ± 0.19	[d]1.60	78 ± 4	This study
[a]Closed DHS	Polyurethane sponge	Anammox granules and activated sludge	35	[c]0.7-1	[c, d]1.94-2.98	[c, d]1.85-2.01	[c]60-95	Chuang *et al.*, 2008
[a]Up-flow column with vertical shaft	Polyethylene sponge strips mounted on a shaft (*Anammox granules accumulated also in the bottom of the reactor*)	Anammox sludge	35 ± 1	2.8	4 ± 0.1	3.6	77	Zhang *et al.*, 2010
[a]Sequencing Batch Biofilm	High-density polyethylene rings	Anammox sludge and anaerobic sludge	35 ± 1	6	1.43 - 1.62	-----	86.8 - 88.5	Yu *et al.*, 2012
[a]Anaerobic upflow fixed bed	Polyurethane sponge	Activated sludge from manure treatment plant	35	40	-----	0.250	85	Monballiu *et al.*, 2013
[b]Integrated fixed-film activated sludge ANITA™ Mox	K5 plastic carriers: AnoxKaldnes (*AOO mainly present as suspended biomass; Anammox principally attached to the plastic carriers*)	Pre-colonized K5 carriers from full scale ANITA™ Mox plant	30 ± 3	20-24	1.8	1.6	80	Veuillet *et al.*, 2014

[a] Lab scale. [b] Full scale. [c] The last 45 days of 330 days of operation. [d] Per m^3 of sponge.

Likely, the high biomass retention capacity of the CSTF reactors provided suitable conditions for the Anammox bacteria for adaptation and acclimatization to 20°C, thereby, minimizing the potential biomass wash-out, irrespective the applied relatively short HRT (of 1.05-1.20 h). Moreover, the nitrogen conversion rate achieved by CSTF-1 at 20°C (1.52 kg-N/m^3sponge·d) was similar to that of the IFAS system at 30 ± 3°C, i.e. 1.6 kg-N/m^3sponge·d (Veuillet *et al.*, 2014) and the closed DHS reactor at 35°C, i.e. 1.85 kg-N/m^3sponge·d (Chuang *et al.*, 2008). Therefore, the implementation of the CSTF might facilitate nitrogen removal in regions with moderate weather conditions. Other characteristics that the CSTF reactors of this research have in common with some trickling filters are the recirculation ratio and the hydraulic loading rate applied. The CSTF reactors operated under a recirculation ratio of 1, which is in the range reported by Daigger *et al.* (2011) for full-scale trickling filters (i.e. 0.5 - 4.0). With regard to the hydraulic loading rate, the value of 13.7 m^3/m^2·d applied in the CSTF of our present research is similar to the loadings reported in small-scale treatment systems (300 inhabitants) that range around 10 - 12 m^3/m^2·d (Almeida *et al.*, 2013) and for full-scale trickling filters designed for carbon oxidation and nitrification, e.g. 14.7 m^3/m^2·d (Daigger *et al.*, 2011).

The design of the CSTF in four sections, and not as one complete sponge bed, avoids the overpressure on the sponge's cubes due to the weight and gives the possibility to have an easy access to the sponges for troubleshooting (e.g. clogging). In the first months, stagnation of liquid was observed in sections 1 and 2 of the CSTF reactors. The presence of a significant growth of biomass was considered to be the cause, but actually an increase in the gas pressure inside the reactor hindered the wastewater flow. To solve this problem, a non-return valve was installed in the upper part of section 1, between the sprinkler system and the sponge layer (Figure 4.1, gas releasing port). Because of the biogas production inside the closed reactors, the gas generated needs to be released to guarantee a stable operation.

The use of the CSTF reactors represents an alternative for nitrogen removal and can contribute to expand the options for the application of Anammox-based technology for wastewater treatment. However, further research is still needed, particularly on the effects of organics on the nitrogen conversions in the CSTF reactors, which could likely come from preceding (anaerobic) treatment systems, including the assessment of the microbial population dynamics and their spatial distribution. Also, it is essential to take into account other aspects like the composition of the biogas generated with particular emphasis on Nitrous Oxide (N_2O) emissions. Nevertheless, the development of a system consisting of an Upflow Anaerobic Sludge Blanket (UASB) reactor, followed by a single stage Anammox sponge-bed trickling filter with limited aeration (achieving partial nitritation and autotrophic nitrogen removal) could be a promising alternative for carbon and nitrogen removal in (sub) tropical climate zones.

4.4. Conclusions

Anammox bacteria were successfully immobilized and cultivated in CSTF systems operated at 20 and 30^{0}C. The temperature of 30^{0}C led to a faster reactor stability and higher removal capacity in a shorter period of time compared to the CSTF operated at 20^{0}C. After an acclimatization and adaptation period of about 45 and 104 days for the CSTF reactors at 30^{0}C and 20^{0}C, respectively, the volumetric nitrogen conversion rate was about 1.52-1.60 kg-N/m$^3_{sponge}$·d with a short HRT of about 1.05-1.20 h and an average total nitrogen removal efficiency of 74 ± 5 % - 78 ± 4 %.

References

Abma W.R., Driessen W., Haarhuis R., van Loosdrecht M.C.M., 2010. Upgrading of sewage treatment plant by sustainable and cost-effective separate treatment of industrial wastewater. Water Science and Technology, 61 (7), 1715-1722.

Almeida P.G.S., Marcus A.K., Rittmann B.E., Chernicharo C.A.L., 2013. Performance of plastic - and sponge - based trickling filters treating effluents from an UASB reactor. Water Science and Technology, 67 (5), 1034-1042.

Chernicharo, C.A.L., van Lier J.B., Noyola A., Bressani Ribeiro T., 2015. Anaerobic sewage treatment: state of the art, constraints and challenges. Reviews in Environmental Science and Bio/Technology, 14 (4), 649-679.

Chuang H.P., Yamaguchi T., Harada H., Ohashi A., 2008. Anoxic ammonium oxidation by application of a down-flow hanging sponge (DHS) reactor. J. Environ. Manage., 18 (6), 409-417.

Daigger G.T., Boltz J.P., 2011. Trickling Filter and Trickling Filter-Suspended Growth Process Design and Operation: A State-of-the-Art Review. Water Environment Research, 83 (5), 388-404.

Dosta J., Fernández I., Vázquez-Padín J.R., Mosquera-Corral A., Campos J.L., Mata-Álvarez J., Méndez R., 2008. Short- and long-term effects of temperature on the Anammox process. Journal of Hazardous Materials, 154, 688-693.

Furukawa K., Lieu P.K., Tokitoh H., Fujii T., 2006. Development of single-stage nitrogen removal using anammox and partial nitritation (SNAP) and its treatment performances. Water Science and Technology, 53 (6), 83-90.

Hendrickx T.L.G., Wang Y., Kampman C., Zeeman G., Temmink H., Buisman C.J.N., 2012. Autotrophic nitrogen removal from low strength waste water at low temperature. Water Research, 46, 2187-2193.

Isaka K., Date Y., Kimura Y., Sumino T., Tsuneda S., 2008. Nitrogen removal performance using anaerobic ammonium oxidation at low temperatures. FEMS Microbiol Lett., 282 (1), 32-38.

ISO 7890/1-1986 (E). Water quality - Determination of nitrate - Part 1: 2,6-Dimethylphenol spectrometric method.

Joss A., Salzgeber D., Eugster J., Koning R., Rottermann K., Burger S., Fabijan P., Leumann S., Mohn J., Siegrist H., 2009. Full-scale nitrogen removal from digester liquid with Partial Nitritation and ANAMMOX in one SBR. Environ. Sci. Technol., 43, 5301-5306.

Lackner S., Horn H., 2013. Comparing the performance and operation stability of an SBR and MBBR for single-stage nitritation-anammox treating wastewater with high organic load. Environmental Technology (United Kingdom), 34 (10), 1319-1328.

Lotti T., Kleerebezem R., Lubello C., van Loosdrecht M.C.M., 2014. Physiological and kinetic characterization of a suspended cell anammox culture. Water Research, 60, 1 -14.

Lydmark P., Lind M., Sörensson F., Hermansson M., 2006. Vertical distribution of nitrifying populations in bacterial biofilms from a full-scale nitrifying trickling filter. Environmental Microbiology, 8 (11), 2036-2049.

Ma B., Pen Y., Zhang S., Wang J., Gan Y., Chang J., Wang S., Wang S., Zhu G., 2013. Performance of anammox UASB reactor treating low strength wastewater under moderate and low temperatures. Bioresource Technology, 129, 606-611.

Metcalf and Eddy, 2003. Wastewater Engineering Treatment and Reuse. McGraw-Hill, New York. USA, pp. 234-238.

Monballiu A., Desmidt E., Ghyselbrecht K., De Clippeleir H., Van Hulle S.W.H., Verstraete W., 2013. Enrichment of anaerobic ammonium oxidizing (Anammox) bacteria from OLAND and conventional sludge: Features and limitations. Separation and Purification Technology, 104, 130-137.

NEN, 1983. Photometric determination of ammonia in Dutch system. In: Nederlandse Normen (Dutch Standards), International Organization for Standardization (ed.) NEN 6472, Dutch Institute of Normalization, Delft, the Netherlands.

Pellicer-Nacher C., Sun S.P., Lackner S., Terada A., Schreiber F., Zhou Q., Smets B.F., 2010. Sequential aeration of membrane-aerated biofilm reactors for high-rate autotrophic nitrogen removal: experimental demonstration. Environ. Sci. Technol., 44 (19), 7628-7634.

Pynaert K., Smets B.F., Beheydt D., Verstraete W., 2004. Start-up of autotrophic nitrogen removal reactors via sequential biocatalyst addition. Environ. Sci. Technol., 38 (4), 1228-1235.

Schmid M., Twachtmann U., Klein M., Strous M., Juretschko S., Jetten M., Metzger J.W., Schleifer K.H., Wagner M., 2000. Molecular evidence for genus level diversity of bacteria capable of catalyzing anaerobic ammonium oxidation. Systematic and Applied Microbiology, 23 (1), 93-106.

Schmid M., Schmitz-Esser S., Jetten M., Wagner M., 2001. 16S-23S rDNA intergenic spacer and 23S rDNA of anaerobic ammonium-oxidizing bacteria: implications for phylogeny and in situ detection. Environ. Microbiology, 3, 450-459.

Schmid M., Walsh K., Webb R., Rijpstra W.I.C., van de Pas-Schoonen K., Verbruggen M.J., Hill T., Moffet B., Fuerst J., Schouten S., Damsté J.S.S., Harris J., Shaw P., Jetten M., Strous M., 2003. Candidatus "Scalindua brodae", sp nov., Candidatus "Scalindua wagneri", sp nov. Two New Species of Anaerobic Ammonium Oxidizing Bacteria. Systematic and Applied Microbiology, 26 (4), 529-538.

Schmid M.C., Maas B., Dapena A., van de Pas-Schoonen K., van de Vossenberg J., Kartal B., van Niftrik L., Schmidt I., Cirpus I., Kuenen J.G., Wagner M., Sinninghe Damsté J.S., Kuypers M., Revsbech N.P., Mendez R., Jetten M.S.M., Strous M., 2005. Biomarkers for in situ detection of anaerobic ammonium-oxidizing (Anammox) bacteria. Applied and Environmental Microbiology, 71 (4), 1677-1684.

Standard Methods for the Examination of Water and Wastewater, 2012, 22 ed. American Public Health Association/American Water Works Association/Water Environment Federation, Washington DC, USA.

Strous M., Heijnen J.J., Kuenen J.G., Jetten M.S.M., 1998. The sequencing batch reactor as a powerful tool for the study of slowly growing anaerobic ammonium-oxidizing microorganisms. Appl. Microbiol., 50 (5), 589-596.

van de Graaf A.A., de Bruijn P., Robertson L.A., Jetten M.S.M., Kuenen J.G., 1996. Autotrophic growth of anaerobic ammonium-oxidizing micro-organisms in a fluidized bed reactor. Microbiology, 142, 2187-2196.

van der Star W.R.L., Abma W.R., Blommers D., Mulder J.W., Tokutomi T., Strous M., Picioreanu C., van Loosdrecht M.C.M., 2007. Startup of reactors for anoxic ammonium oxidation: Experiences from the first full-scale anammox reactor in Rotterdam. Water Research, 41, 4149-4163.

van Hulle S.W.H., Vandeweyer H., Audenaert W., Monballiu A., Meesschaert B., 2011. Influence of the feeding regime on the start-up and operation of the autotrophic nitrogen removal process. Water SA, 37, 289-294.

Veuillet F., Lacroix S., Bausseron A., Gonidec E., Ochoa J., Christensson M., Lemaire R., 2014. Integrated fixed-film activated sludge ANITA^{TM} Mox process - a new perspective for advanced nitrogen removal. Water Science and technology, 69 (5), 915-922.

Wett B., Hell M., Nyhuis G., Puempel T., Takacs I., Murthy S., 2010. Syntrophy of aerobic and anaerobic ammonia oxidisers. Water Science and Technology, 61 (8), 1915-1922.

Wilsenach J., Burke L., Radebe V., Mashego M., Stone W., Mouton M., Botha A., 2014. Anaerobic ammonium oxidation in the old trickling filters at Daspoort Wastewater Treatment Works. Water SA, 40 (1), 81-88.

Yu Y.C., Gao D.W., Tao Y., 2012. Anammox start-up in sequencing batch biofilm reactors using different inoculating sludge. Appl. Microbiol. Biotechnol., 97 (13), 6057-6064.

Zhang L., Yang J., Ma Y., Li Z., Fujii T., Zhang W., Takashi N., Furukawa K., 2010. Treatment capability of an up-flow anammox column reactor using polyethylene sponge strips as biomass carrier. Journal of Bioscience and Bioengineering, 110 (1), 72-78.

Autotrophic nitrogen removal over nitrite in a sponge-bed trickling filter under natural air convection

Contents

This chapter is based on: Sánchez Guillén J.A., Jayawardana L.K.M.C.B., Lopez Vazquez C.M., de Oliveira Cruz L.M., Brdjanovic D., van Lier J.B., 2015. Autotrophic nitrogen removal over nitrite in a sponge-bed trickling filter. Bioresource Technology, 187, 314-325.

Abstract

Partial nitritation in sponge-bed trickling filters (STF) under natural air circulation was studied in two reactors: STF-1 and STF-2 operated at 30°C with sponge thickness of 0.75 and 1.50 cm, respectively. The coexistence of AOO and Anammox bacteria was obtained and attributed to the favorable environment created by the reactors' design and operational regimes. After 114 days of operation, the STF-1 had an average NH_4^+-N removal of 69.3% (1.17 kg $N/m^3_{sponge} \cdot d$) and a total nitrogen removal of 52.2% (0.88 kg $N/m^3_{sponge} \cdot d$) at a Nitrogen Loading Rate (NLR) of 1.68 kg $N/m^3_{sponge} \cdot d$ and Hydraulic Retention Time (HRT) of 1.71h. The STF-2 showed an average NH_4^+-N removal of 81.6 % (0.77 kg $N/m^3_{sponge} \cdot d$) and a total nitrogen removal of 54% (0.51 kg $N/m^3_{sponge} \cdot d$), at an NLR of 0.95 kg $N/m^3_{sponge} \cdot d$ and HRT of 2.96 h. The findings suggest autotrophic nitrogen removal over nitrite in STF.

5.1. Introduction

Upflow Anaerobic Sludge Blanket (UASB) reactors have been recognized as suitable sewage treatment processes in developing countries because of their low energy use, easy maintenance and cost-effectiveness. Hundreds of full-scale examples can be found worldwide for sewage treatment, particularly in Latin America and India (van Lier *et al.*, 2010; Chernicharo *et al.*, 2012, 2015). However, anaerobic wastewater treatment systems only target carbonaceous compounds leaving nutrients in their effluents. The latter may become cumbersome when more strict discharge standards are applied to the receiving water bodies. In those cases, suitable post-treatment processes are required to be installed after UASB treatment.

Anammox bacteria have opened the possibility to achieve cost-effective biological nitrogen removal from municipal wastewater when coupled to UASB reactors. Nevertheless, an important technical challenge to sustain the Anammox process is the required partial nitritation to supply ammonium (NH_4^+-N) and nitrite (NO_2^--N) in the appropriate ratio to the biomass. Moreover, for the envisaged application, the foreseen technology should not depend on mechanical aeration, as this will add to the complexity and costs.

Results from previous research described in Chapter 4, showed proved that Anammox bacteria could be sustained in closed sponge-bed trickling filters (CSTFs). The next step for developing an appropriate low cost reactor system for N removal is including partial nitritation by ammonia oxidizing organisms (AOO) into the sponge-bed biomass. Uemura *et al.* (2011) reviewed that, so far, partial nitritation research is focused on (i) reactors operated at 30-40^{0}C, (ii) reactors operated at high salinity concentrations, (iii) the application of mechanically controlled low dissolved oxygen (DO) concentrations, (iv) high inorganic carbon content effluents, (v) the inhibition of nitrite oxidizing organisms (NOO) by free ammonia or free nitrous acid, and (vi) a combination of the previous factors. A thorough list of full-scale facilities achieving partial nitritation, by means of some of the above parameters, and the Anammox process for side-stream treatment is described elsewhere (Lackner *et al.*, 2014).

The down-flow hanging sponge (DHS) reactor has been developed for post-treatment of UASB effluents as an affordable, easy-maintenance and promising sewage treatment for developing countries (Machdar *et al.*, 2000; Tandukar *et al.*, 2006). The DHS reactor is a sponge-based trickling filter that uses polyurethane sponge that hangs freely in the air as support media to retain biomass. The oxygen in the air is dissolved into the trickling wastewater from the top of the reactor, providing the required DO for the growth of microorganisms retained both inside and outside the sponge media. Therefore, no external aeration is needed. Furthermore, Tandukar *et al.* (2006) demonstrated that the UASB-DHS system has produced excellent results to remove COD, BOD and to some extent nitrogen from sewage.

By applying a closed DHS reactor for NH_4^+ rich synthetic wastewater (100 mg-NH_4^+-N/L) under controlled oxygen conditions using an air pump, Chuang *et al.* (2007) attained partial nitritation achieving about 50% ammonium conversion to nitrite at 0.2 mg/L of DO and 30°C. Though promising, this application requires mechanical control equipment to control the DO in the reactor, and therefore elevated capital investment, high operational costs and advanced technical expertise.

On the other hand, the use of a sponge-based trickling filter unit with natural air convection, being the DHS reactor the major and most important example of such system, could be a promising and cost-effective approach as a post- treatment step for ammonium-rich UASB effluents, in which partial nitritation and the Anammox process could developed. In this regard, Machdar *et al.* (2000) observed in a DHS reactor a DO gradient from 7.5 mg-O_2/L in the external layers of the sponge to around 0.2 mg-O_2/L in the inner layers (1 cm inside the sponge). Such DO conditions may be favorable to sustain partial nitritation in sponge-bed filters (Chuang *et al.*, 2007).

Moreover, sponge-based trickling filters possess other advantageous properties such as (i) a large surface area that can lead to an increased biomass retention capacity, thus being able to attain long solids retention times (SRT) favorable for slow growing organisms, (ii) potentially high microbial conversions as a consequence of the high biomass retention and high permeability, which could be reflected in shorter hydraulic retention times, (iii) presumably low construction costs and low space requirements, and (iv) low operational costs, since no mechanical aeration and less complicated control equipment would be required.

This research aims to assess the feasibility to attain partial nitritation with natural air convection in lab-scale sponge bed trickling filters as a low cost post-treatment step using a synthetic substrate that simulates an ammonium rich effluent (100 mg of NH_4^+-N/L) from a UASB system treating municipal wastewater at 30°C. Suitable operational parameters, such as sponge thickness, NLR and HRT, are also explored towards the development of a cost-effective autotrophic nitrogen removal process over nitrite in sponge bed trickling filter systems.

5.2. Materials and methods

5.2.1. Design of the reactors

The partial nitritation was carried out in two lab-scale flow-through type Sponge-bed Trickling Filter (STF) units (namely STF-1 and STF-2). The trickling filter units were constructed using transparent acrylic glass. Horizontally layered polyurethane sponge slabs BVB Sublime (second generation) (BVB Substrates; De Lier; the Netherlands) were used as biomass support media.

The thickness of the sponge support material was chosen taking into account the studies of Araki, *et al.* (1999) who demonstrated that the sponge material maintains aerobic conditions down to the depth of 0.75cm from the surface, beyond which is an anoxic environment, and the studies of Machdar *et al.,* (2000) who observed a DO concentration of 0.2 mg-O_2/L at 1.0 cm inside the sponge from surface. Thus, the sponge sheet thickness was 0.75 and 1.50 cm for STF-1 and STF-2, respectively. The configuration of the two reactors is summarized in Table 5.1 and Figure 5.1 illustrates a schematic diagram of each reactor.

Both reactors were operated with natural air convection at 30°C in a temperature-controlled room. Air circulation across sponge medium was facilitated through lateral openings located above each sponge layer. All openings (diameter of 4 mm) were kept open at the start-up phase to ensure nitritation and, in a step wise manner, gradually closed in order to reach low dissolved oxygen (DO) concentrations, i.e. below 2 mg/L, across certain sponge layers to attain partial nitritation.

In some occasions during the experimental period, the mentioned lateral openings were partly re-opened to provide additional oxygen. To limit air circulation over the height of the reactor, the sponge layers had the same cross-sectional area as the reactor's surface area. Synthetic wastewater was fed from the top of each reactor with a miniature water distributor (shower). To minimize the influent DO concentrations, the demineralised water, by far having the largest share of the synthetic wastewater, was periodically flushed with nitrogen gas. In addition, an oxygen scavenger water lock, containing sodium sulphite and cobalt chloride, was installed in the ventilation located on top of the demineralized water tank.

Table 5.1. Configuration of the two lab scaled Sponge-bed Trickling Filters.

Parameter (unit)	STF-1	STF-2
Overall Reactor height (cm)	60.5	46.0
Effective Reactor height (cm)	54	39
Reactor shaft size - internal (cm^2)	6.75 x 6.75	6.75 x 6.75
Sponge sheet size (cm^2)	6.75 x 6.75	6.75 x 6.75
Sponge sheet thickness (cm)	0.75	1.50
Sponge void ratio (%)	98	98
Sponge density (kg/m^3)	28	28
Number of sponge layers	29	15
Total sponge volume (cm^3)	991	1025
Specific surface area (m^2/m^3)[a]	326	193
Spacing between sponge layers (cm)	1.0	1.0
Volume Fraction of sponge medium (%)	37	47

[a]Based on the total number of sponge sheets per reactor, the surface area of each sponge sheet and the total volume of sponge sheets.

5.2.2. Synthetic substrate

Synthetic substrate was composed of two solutions divided in two containers (Figure 5.1): the ammonium feed container and the bicarbonate feed container. The composition of these substrates was modified from van de Graaf *et al.*, 1996, and contained per 1 liter of demineralized water in the ammonium-rich feed: 5.9656 g NH_4Cl; 0.77 g $MgSO_4 \cdot 7H_2O$; 0.3906 g KH_2PO_4; 4.6875 g $CaCl_2 \cdot 2H_2O$ and in the bicarbonate feed: 0.1786 g $FeSO_4 \cdot 7H_2O$; 19.531 g $KHCO_3$; 0.1786 g NaEDTA and 1.25 mL of trace element solution. The trace element solution contained per liter: 15 g EDTA; 0.43 g $ZnSO_4 \cdot 7H_2O$; 0.24 g $CoCl_2 \cdot 6H_2O$; 0.99 g $MnCl_2 \cdot 4H_2O$; 0.25 g $CuSO_4 \cdot 5H_2O$; 0.22 g $Na_2MoO_4 \cdot 2H_2O$; 0.19 g $NiCl_2 \cdot 6H_2O$; 0.1076 g Na_2SeO_4; 0.014 g H_3BO_3; 0.05 g $NaWO_4 \cdot 2H_2O$. After mixing the solutions with demineralized water, the NH_4^+-N concentration in the synthetic substrate was approximately 100 mg/L with a pH of about 7.8.

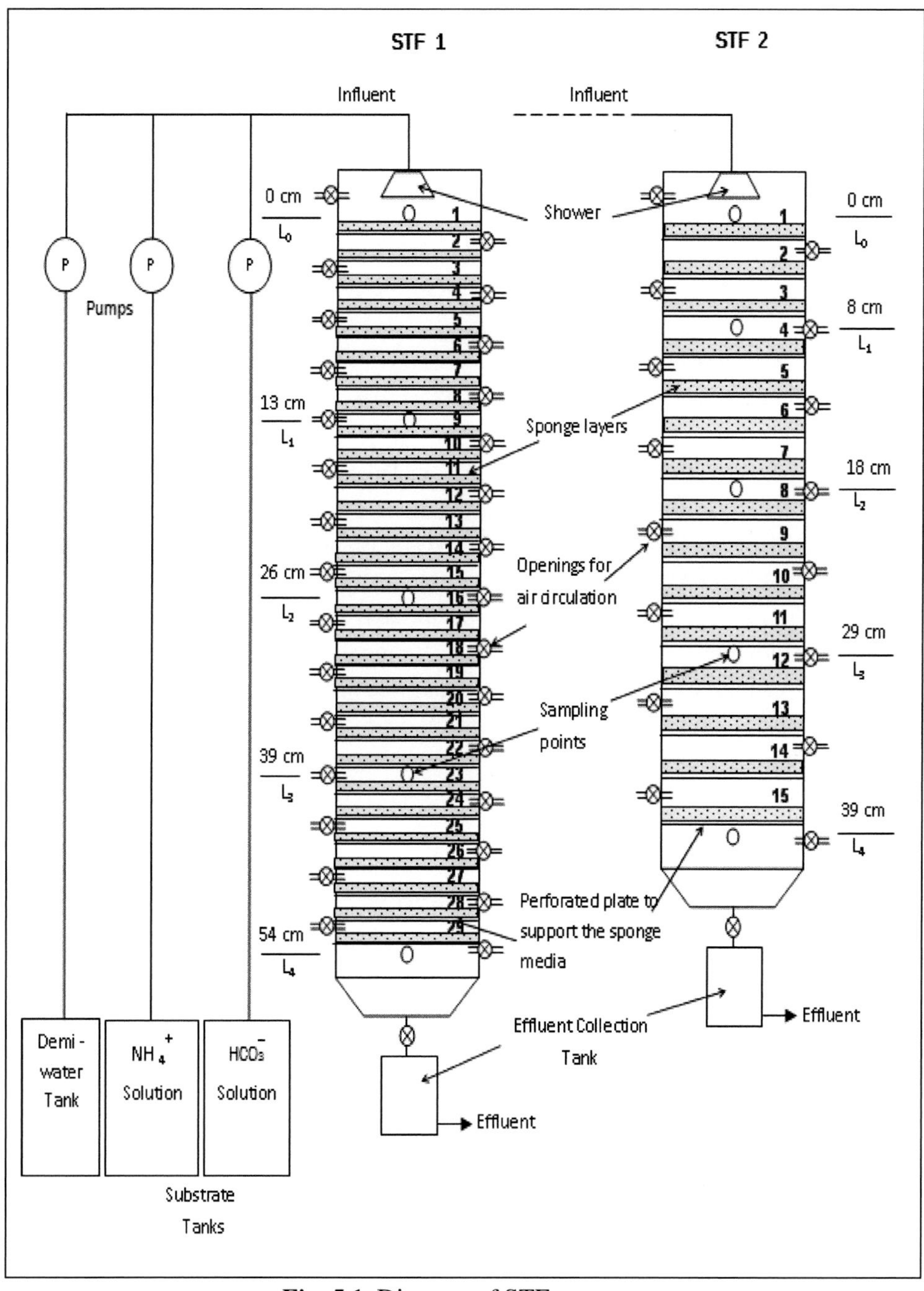

Fig. 5.1. Diagram of STF reactors.

5.2.3. Experimental phases

5.2.3.1. Inoculation

Activated sludge from the Harnaschpolder wastewater treatment plant (Den Hoorn, The Netherlands) was used for the inoculation of STF reactors. The sponge layers were properly saturated with synthetic wastewater and the activated sludge was aerated during 10 minutes prior to inoculation. The sponge layers were seeded from top to bottom using the lateral openings of each reactor. All sponge layers were inoculated with approximately 60 mL of activated sludge with a Mixed Liquor Suspended Solids (MLSS) content of 4,352 mg/L (STF-1) and 3,725 mg/L (STF-2).

5.2.3.2. Phase I: Start-up

In the start-up phase, a constant influent flow rate of 5.7 L/d was fed into both reactors. The average Nitrogen Loading Rate (NLR) for the entire period was 0.69 (STF-1) and 0.66 (STF-2) kg-NH_4^+-N/m^3_{sponge}·d. This phase continued for 45 days and 58 days for STF-1 and STF-2, respectively. During Phase I, the nitrogen conversions were assessed by analysing the NH_4^+- N, NO_2^- - N and NO_3^- - N concentrations at different levels including the influent and effluent of both reactors to follow up their stability. The nitrification activity achieved in each reactor was used as a decision criterion to move on to the next phase, i.e. the increase in NLR to enhance partial nitritation. In addition, Total Suspended Solids (TSS) analyses were carried out in the effluent every 5 days and tracer tests were done to assess the actual Hydraulic Retention Time (HRT) of the STF reactors. Fluorescent in situ hybridization (FISH) analyses were conducted to identify the organisms present in the biofilms.

5.2.3.3. Phase II: Increment of NLR and adjustment of air inlet points

In Phase II, the NLR was increased in a stepwise manner by adjusting the feeding rate of both reactors while maintaining the influent concentrations. During this phase, the NLR was increased up to average values of 1.64 kg-NH_4^+-N/m^3_{sponge}·d in STF-1 and 0.91 kg-NH_4^+-N/ m^3_{sponge}·d in STF-2. The air circulation through the reactors was adjusted from time to time based on the nitritation/nitrification performance aiming to achieve successful partial nitritation. In addition, soluble Chemical Oxygen Demand (COD) and effluent TSS concentrations were also measured. A tracer test was carried out to assess the actual HRT at the final stage of the experiment. A summary of the operational phases for both reactors and the average values for each parameter are listed in Table 5.2 and Table 5.3.

Table 5.2. Details of operational phases and parameters for the Sponge-bed Trickling Filter 1 (STF-1).

Reactor	Parameter (unit)	Phase I	Phase II			
		Start-up	Phase II-A[a]	Phase II-B	Phase II-C	Phase II-D
STF-1	Phase started date after inoculation (d)	-	45	51	63	80
	Duration (d)	45	6	12	17	33
	Flow (L/d)	5.7	11.2	14.9	14.9	14.9
	Theoretical HRT (h)[b]	4.2	2.4	1.6	1.6	1.6
	Hydraulic Loading Rate (HLR; $m^3/m^2 \cdot d$)[c]	1.3	2.2	3.3	3.3	3.3
	NH_4^+-N (mg/L)	111.7 ± 2.5	111.5	112.5 ± 0.8	107.1 ± 2.1	111.9 ± 5.5
	NLR[d] (kg-NH_4^+-N/$m^3_{sponge} \cdot d$)	0.64 ± 0.01	1.26	1.69 ± 0.01	1.61 ± 0.03	1.68 ± 0.08
	Air inputs to sponge layers (number of layers)	Opened (29)	Opened (29)	Opened (29)	Additional air inputs (20), totally opened	No air input to 24% of sponge layers (7)

[a] Intermediate phase. [b] Based on sponge volume. [c] Theoretical value. [d] Average values at stable conditions.

Table 5.3. Details of operational phases and parameters for the Sponge-bed Trickling Filter 2 (STF-2).

Reactor	Parameter (unit)	Phase I	Phase II			
		Start-up	**Phase II-A**[a]	**Phase II-B**	**Phase II-C**	**Phase II-D**
STF-2	Phase started date after inoculation (d)	-	46	60	81	91
	Duration (d)	46	14	21	10	23
	Temperature (ºC)	30	30	30	30	30
	Flow (L/d)	5.7	5.7	8.2	8.2	8.2
	Theoretical HRT (h)[b]	4.3	4.3	3.0	3.0	3.0
	Hydraulic Loading Rate (HLR; m^3/m^2 d)[c]	1.3	1.3	1.8	1.8	1.8
	NH_4^+-N (mg/L)	119 ± 13.2	117.3 ± 6.8	106.7 ± 4.8	114.3 ± 3.7	118.5 ± 4.6
	NLR[d] (kg-NH_4^+-N/m^3_{sponge}·d)	0.66 ± 0.07	0.65 ± 0.04	0.85 ± 0.04	0.91 ± 0.03	0.95 ± 0.04
	Air inputs to sponge layers (number of layers)	Opened (15)	Additional air inputs (15), totally opened	Totally opened	No air input to 27% of sponge layers (4)	Totally opened

[a] Intermediate phase. [b] Based on sponge volume. [c] Theoretical value. [d] Average values at stable conditions.

5.2.4. Tracer test

The step method described by Levenspiel (1999) was used to determine the actual HRT of both reactors during Phase I. A lithium chloride (LiCl) solution was added continuously as tracer to the influent demineralised water line. The Li-solution was mixed with the substrate flow reaching a reactor influent concentration of around 10 mg-Li/L. Samples were collected in intervals of 15 minutes in the effluent of each reactor and immediately preserved by acidification (by HNO_3 addition to pH 1-2). In the final stage of Phase II, the pulse method was used to determine the HRT of both reactors. Thus, 50 mL of a 10 mg-LiCl/L solution were added as a spike to each reactor. Effluent samples were taken in intervals of 10 minutes and immediately acidified. An atomic absorption spectrophotometer AANALYST 200 (Perkin Elmer, USA) was used for the analysis of lithium concentrations in the effluents. The lithium concentrations as function of time were used to calculate the exit age or E function and thereby to determine the residence time distribution curve, where the area under the curve represents the average residence time of the each reactor.

5.2.5. Anammox activity tests

At the end of Phase II, oxygen was excluded in the STF reactors to assess the anaerobic ammonium oxidation activity. All the air inlet points were closed using non-return air valves (check valves) that allowed the air located inside the reactors to flow out, while avoiding air intrusion. Both STF were sparged with nitrogen gas for 3 to 4 hours prior to the execution of the tests to create anaerobic conditions inside the reactors. The substrate bottles and demineralised water tanks were also flushed with nitrogen gas reaching a final DO concentration of less than 0.1 mg/L. Tedlar bags of 10 L filled with nitrogen gas were connected to the substrate bottles to avoid the generation of a negative pressure inside the reactors. Thereafter, approximately, 50 mg of NH_4^+- N/L and 50 mg of NO_2^- - N/L were continuously fed to STF-1 and STF-2 for 3 and 4 hours, respectively, prior to sampling. Thereafter, every hour samples were collected for the determination of nitrite and ammonium consumption and nitrate production.

5.2.6. Analytical methods

The TSS, COD, alkalinity and nitrite-nitrogen (NO_2^--N) concentrations were determined according to Standard Methods for the Examination of Water and Wastewater (2012). Effluent wastewater collected over a day from each reactor was well mixed in order to have representative samples (100 mL) for TSS analysis.

Ammonium-nitrogen (NH_4^+-N) was measured spectrophotometrically following the standard NEN 6472 (NEN, 1983). For the nitrate-nitrogen (NO_3^--N) measurement the standard ISO 7890-1:1986 (ISO 7890/1, 1986) was applied. pH was measured with a portable pH meter (Model ProfiLine 3310. WTW, Germany) and the DO concentrations were recorded using a portable DO meter (Model pH/Oxi 340i/3400i. WTW, Germany). For pH and DO measurements, a syringe with a needle was used to collect 10 ml of wastewater samples from sponge layers.

5.2.7. FISH analyses

Fluorescence *in Situ* Hybridization (FISH) technique was carried out to identify the microorganisms located in the biofilms of the sponge layers. Combined biomass samples were scratched out from the sponges located between levels L_0 and L_2 (Figure 5.1). Collected biomass samples were fixed in paraformaldehyde (4%), and hybridizations with fluorescent probes were performed as per the procedure described by Schmid *et al.* (2000). Epifluorescence was used for identification of Anammox bacteria, ammonia-oxidizing organisms (AOO) as well as nitrite-oxidizing organisms (NOO) cells and 4', 6'-diamidino-2-phemylindol (DAPI) as general DNA stain. All gene probes were labelled with either the fluorophores Cy3 or FLUOS (Biomers.net, Germany). An epifluorescence microscope BX51 with a camera XM10 (Olympus, Japan) was used together with the standard software package delivered with the instrument (Version 1.2) for image acquisition.

5.3. Results and discussion

5.3.1. Biomass development

No significant biomass washout was observed during the operation of the STF reactors. For Phase I, the effluent TSS concentrations were 1.58 mg/L ± 1.02 (STF-1) and 1.71 mg/L ±1.17 (STF-2). Similarly, Phase II showed low values of TSS in the effluents, 4.26 ± 0.80 mg/L (STF-1) and 5.36 mg/L ± 0.10 (STF-2). In fact, a remarkable feature of the sponge based bioreactors, e.g. the DHS reactors, is that the excess sludge production rate is very low compared to other treatment systems as activated sludge; 0.09 *vs.* 0.88 g-TSS/g-$COD_{removed}$, respectively (Onodera *et al.*, 2013).

The biomass development in the sponge medium was visually inspected periodically. The biomass growth and expansion over the sponge sheets between levels L_0 and L_2

were higher than that in the other segments in both reactors. This could be because of the direct availability of substrate at the upper sponge layers. After increasing the number of air inlet points a faster growth of biomass was observed that covered the sponge layers and side walls. A gradual growth of reddish biomass was first observed in the sponges of STF-2, mainly at the end of the Start-up phase, and later on noticed in STF-1, especially at the sponges located between levels L_0 and L_2 in both reactors. The distance between the top part of L_0 and L_2 in STF-1 represents about 48% (26 cm) of the effective reactor height (54 cm), whereas in STF-2; it represents about 46% (18 cm) of the effective reactor height (39 cm).

The reddish biomass was located starting from the corners of the reactor cross sections to the middle of each sponge layer and in the interface of the perforated aluminium plates. Likely, in those locations anoxic zones (with low DO concentrations) favourable to Anammox bacteria were created. In fact, there are several examples of bioreactors with support material that were not explicitly designed for Anammox cultivation, where the presence and activity of Anammox bacteria have been observed (Barana *et al.*, 2013; Wilsenach *et al.*, 2014). Subsequently, FISH analyses were carried out and results showed the presence of AOO, NOO and Anammox bacteria (Figure 5.2).

5.3.2. Performance of STF-1

The NH_4^+-N removal performance of STF-1 under different operational parameters and conditions is shown in Figure 5.3. During Phase I, STF-1 was operated at an actual $HRT_{sponge\ volume\ based}$ of 4.60 h (only 9.13 % higher than the theoretical $HRT_{sponge\ volume\ based}$). All the air inlet points were kept open for proper air circulation. After 6 days, the NH_4^+-N removal efficiency was 50% higher. This result was a reflection of the fast growth of AOO in STF-1. In the last 22 days of Phase I, an average nitrogen deficit (nitrogen removal) of 4.3% was observed based on nitrogen mass balances.

In Phase II-A, the influent NLR was gradually increased from 0.64 kg-N/m^3_{sponge}·d to 1.26 kg-N/m^3_{sponge}·d by increasing the substrate flow rate from 5.7 L/d to 11.2 L/d. The NH_4^+-N removal efficiency (i.e. nitrification performance) dropped from 59% to 31.7% in the last days of the Start-up and II-A phases. This phase was limited to 6 days because it was considered as an intermediate phase towards the target NLR of 1.6 kg-N/m^3_{sponge}·d to achieve partial nitritation under limited oxygen condition.

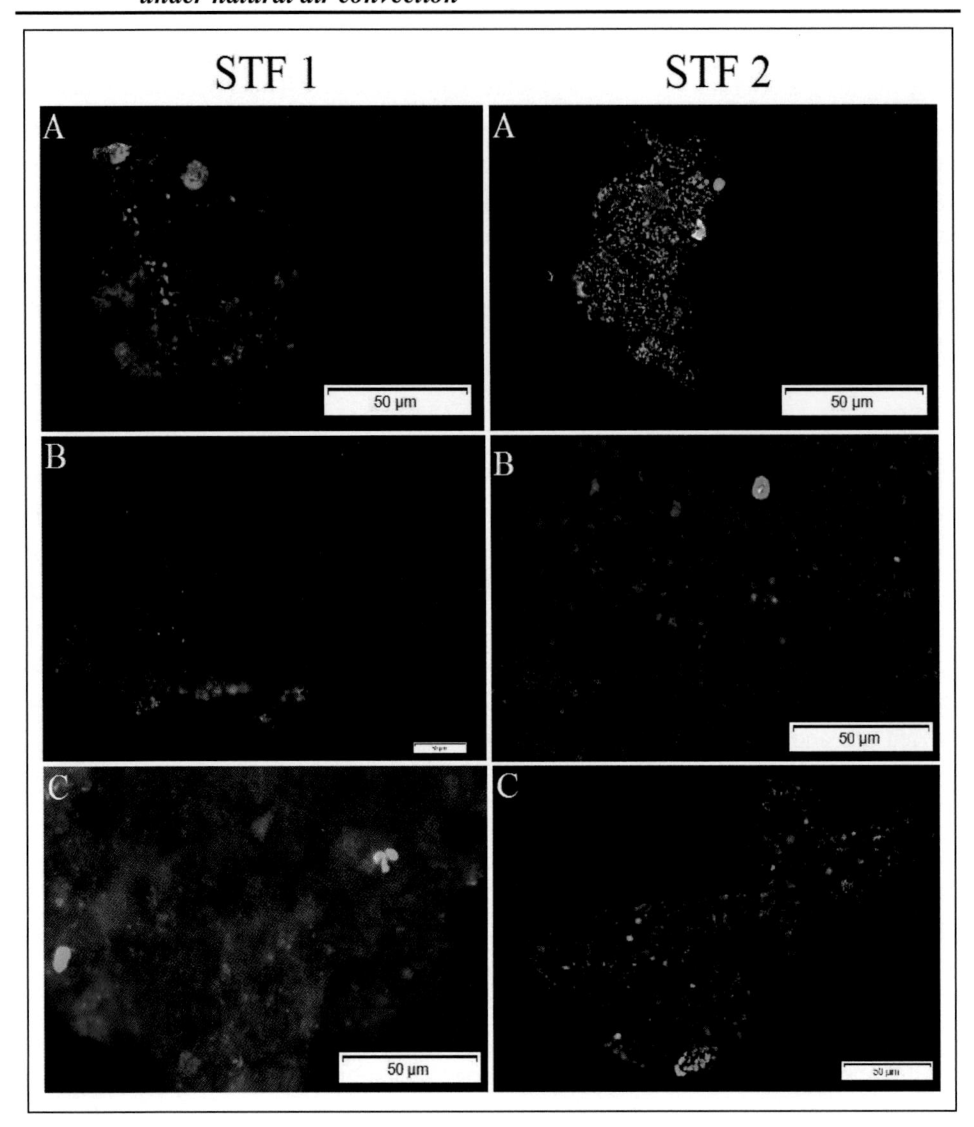

Figure 5.2. Identification by Fluorescence *in situ* Hybridization (FISH) of dominant microbial populations present in layers L_0 to L_2 in the STF reactors. All scale bars are 50 µm and the blue color indicates DAPI as general DNA stain. (**A**) The green color indicates ammonia-oxidizing organisms hybridized with NSO190 probe and red color indicates Anammox bacteria hybridized with AMX 820 probe; (**B**) The green color indicates nitrite oxidizing organisms hybridized with NIT 3 probe and red color indicates Anammox bacteria hybridized with AMX 820 probe; (**C**) The green color indicates ammonia-oxidizing organisms hybridized with NSO190 and NSO1225 probes and red color indicates nitrite oxidizing organisms with Nit 3, NSM 156 and Ntspa 662 probes.

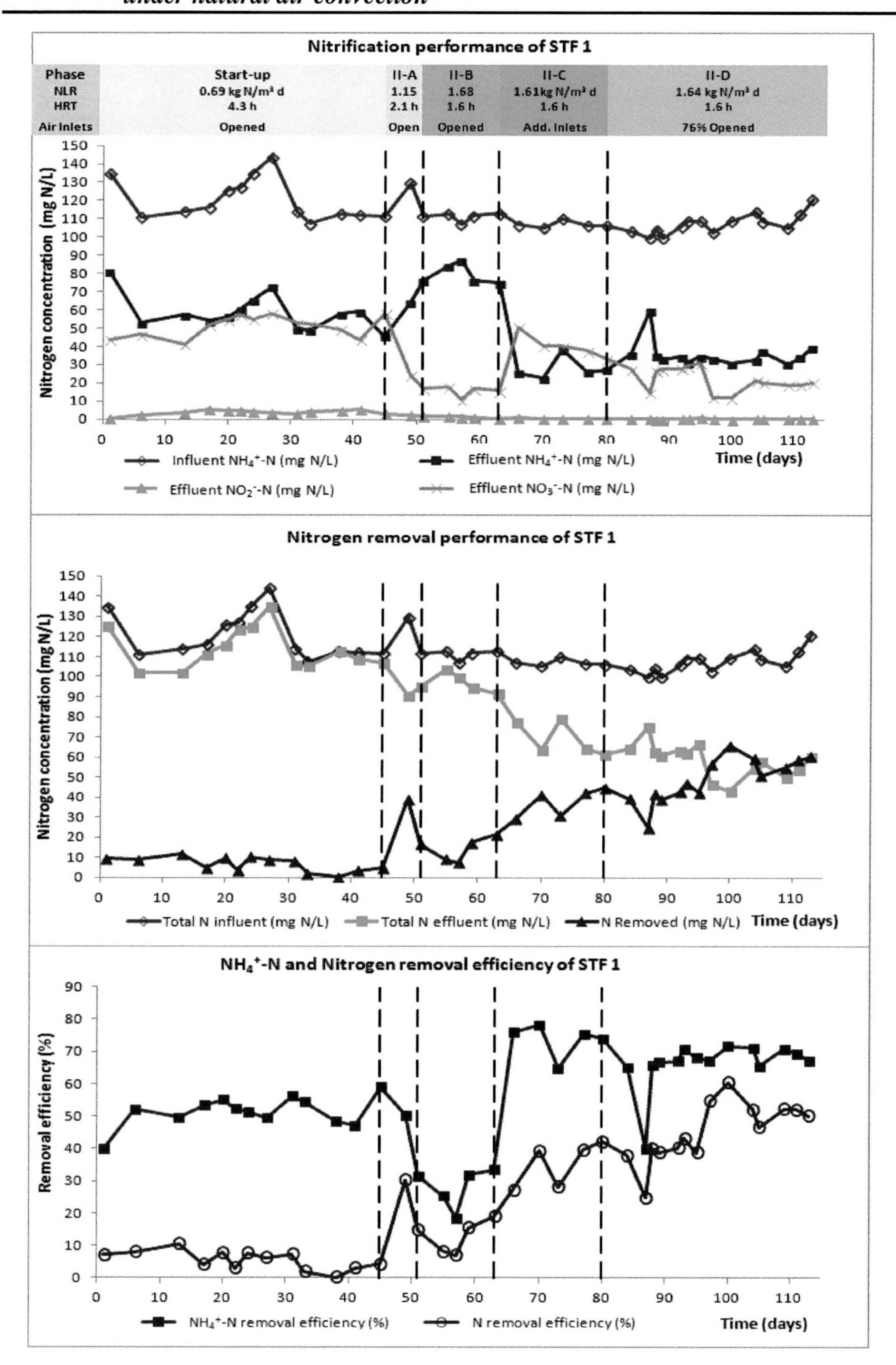

Figure 5.3. NH_4^+-N removal performance of STF-1.

Influent NLR was then further increased to 1.69 kg $N/m^3_{sponge}{\cdot}d$ by increasing the substrate flow rate up to 14.9 L/d in Phase II-B. However, the NH_4^+-N removal efficiency dropped further to 18.5% and gradually recovered leaving 33.6% at the end of this phase. Interestingly, a nitrogen deficit was also observed, which suggested that nitrogen removal was taking place. It was noticed that the nitrogen removal efficiency increased from 8.2% to the maximum of 19% during this phase.

Since there was no improvement in nitritation/nitrification, in the Phase II-C, 20 additional air inlets points were provided at the back side of SFT 1; one opening for each sponge up to the 2/3th depth of the reactor to enhance the reactor's performance while keeping the same substrate flow rate of 14.9 L/d and an NLR of 1.61 kg-$N/m^3_{sponge}{\cdot}d$. The decision of providing additional air inlet points was justified as the reactor quickly responded to the increased air circulation with an increase in the NH_4^+ - N removal efficiency from 33.6% to 76% within 3 days. During the last 11 days of Phase II-C, the average NH_4^+-N removal was 73.1%. The percentage of the effluent nitrate concentration with respect to influent NH_4^+-N concentration accounted for 31.5% at the last day. This was higher in this phase compared to Phase II-A (15.4%) and Phase II-B (14.1%). Furthermore, at the end of Phase II-C, 42.1% nitrogen removal efficiency was achieved.

To reduce the effluent nitrate concentrations, it was decided to close all air inlet points of 7 sponge sheets in STF-1 during Phase II-D. It was postulated that by controlling the DO concentration the nitrification process would be limited and the reaction would stop at the nitritation level. The air inlet points were closed between the sponge layers: 4, 7, 10, 13, 16, 19 and 25 (Figure 5.1), leading to a considerable drop in the nitrate production and a concomitant decrease in NH_4^+-N removal efficiency from 75% to 70%. Nevertheless, the nitrogen removal percentage increased significantly from 42.1% to a maximum of 60.4% during the final part of this phase. Furthermore, for Phase II-D the actual $HRT_{sponge\ volume\ based}$ was 1.71 h, only 6.43% higher than the theoretical $HRT_{sponge\ volume\ based}$.

5.3.3. Performance of STF-2

Figure 5.4 shows the overall NH_4^+-N removal performance of STF-2 under the operational parameters and conditions applied. In the Start-up phase, the STF-2 was operated at a low ammonium loading rate (NLR) of 0.66 kg-$N/m^3_{sponge}{\cdot}d$ at an actual $HRT_{sponge\ volume\ based}$ of 4.12 h (5.58% less than the theoretical $HRT_{sponge\ volume\ based}$).

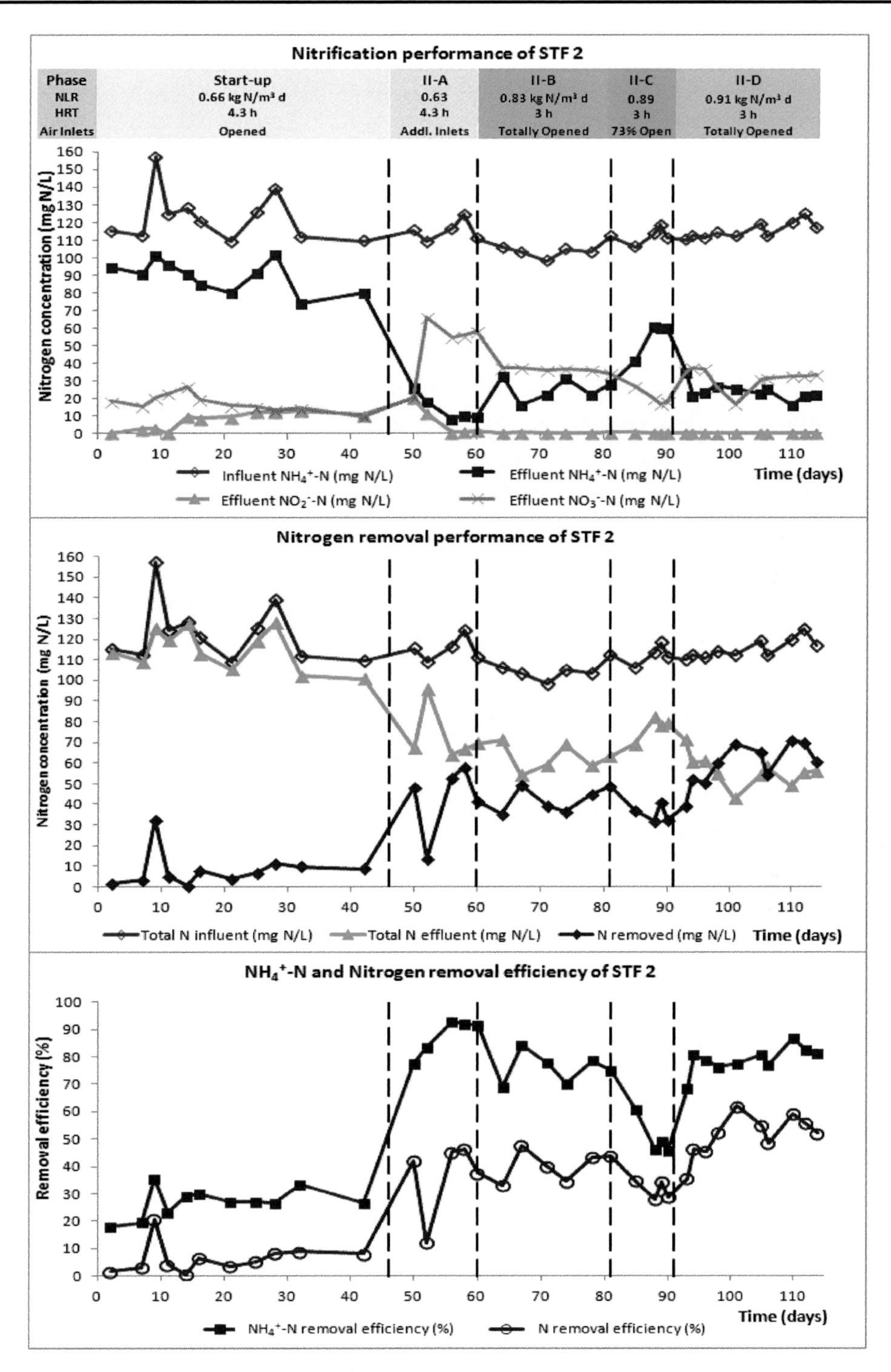

Figure 5.4. NH_4^+-N removal performance of STF-2.

Similar to the STF-1, all the air inlet points at the STF-2 were kept open for proper air circulation. After 9 days around 30% of NH_4^+-N removal efficiency was observed which indicated that AOO were not very active in STF-2 compared to STF-1 during this phase. However, nitrification performance was constant towards the end of Start-up phase. An average nitrogen deficit of 6.7% based on the nitrogen mass balance was observed in STF-2 at the second half of this phase (last 22 days).

Since there was no significant improvement in the STF-2 in terms of nitritation/nitrification during the start-up period, 15 additional air inlet points were created for each sponge level at the back side of the reactor in Phase II-A to enhance the performance, while keeping the same substrate flow rate of 5.7 L/d and a NLR of 0.65 kg-N/m^3_{sponge}·d. Subsequently, during this phase, the NH_4^+-N removal efficiency i.e. nitrification performance significantly increased from 26.7% to 92.8% and remained around this removal efficiency until the end of the phase (at approximately 91.6%). In the meantime, the nitrate production also improved from 8.9% to 52.5%. Furthermore, a significant increase in the nitrogen deficit was noticed from 8% to 37.5% as per the last days of the Start-up and II-A Phases.

In Phase II-B, the influent NLR was increased to 0.85 kg-N/m^3_{sponge}·d by increasing the substrate flow from 5.7 L/d to 8.2 L/d. The NLR increase was based on the observed stable nitrification performance in Phase II-A. During Phase II-B, the NH_4^+-N removal efficiency considerably dropped by 15% compared to the previous phase. Despite this decrease, the nitrogen deficit gradually increased to 43.6% at the end of this phase. To limit the DO concentration, and thus hindering the nitrification process pursuing the reaction to stop at the nitritation level, it was decided to close all air inlets of 4 sponge sheets (sponge number 4, 7, 10 and 13; Figure 5.1) during Phase II-C. However, the reactor's performance dropped further and the NH_4^+-N removal efficiency dropped from 74.8% to 45.7% within 4 days. In addition, the nitrogen removal (deficit) decreased to 28.9%.

In Phase II-D, all the closed air inlet points were re-opened to facilitate a higher air circulation through the reactor. Interestingly, the reactor responded to the change quickly and achieved a NH_4^+-N removal efficiency as high as 86.7%. No accumulation of NO_2^- in the effluent was observed, whereas nitrogen removal increased to a maximum of 61.9% after 101 days of inoculation. In addition, the actual HRT of Phase II-D was close to the theoretical value, 2.96 *vs.* 3.00, correspondingly (1.35 % lesser).

5.3.4. Anammox activity tests

Since no external COD source was supplied, the nitrogen deficit observed in both reactors was attributed to either (i) heterotrophic denitrification using any soluble COD generated by soluble microbial products (SMP) or biomass decay or (ii) autotrophic denitrification by Anammox bacteria. If heterotrophic denitrification would have been occurred, 121 or 203 mg-COD/L would have been required to be produced by SMP or biomass decay to denitrify via nitrite or nitrate reduction, respectively, the 71 mg-N/L of nitrogen removed in STF-2 during Phase II-D (based on 1.71 mg-COD/mg-NO_2^--N and 2.86 mg-COD/mg-NO_3^--N; Ekama and Wentzel, 2008), which seemed unlikely given the reactor's conditions. To validate the second hypothesis, Anammox activity tests were performed in the reactor, following nitrogen conversion stoichiometry as proposed by Lotti *et al.* (2014). The results obtained are tabulated in Table 5.4 and Table 5.5. Figure 5.5 illustrates the nitrogen profiles observed in the anammox activity tests performed in STF-1 for 5 h and in STF-2 for 6 h as well as the nitrogen removal efficiencies along the respective reactor heights.

Despite that the NLR were different for both reactors, they showed the same behaviour in terms of Anammox activity and nitrogen removal. According to the Anammox activity tests, both reactors have over 80% autotrophic nitrogen removal capacities. Consequently, it can be concluded that the nitrogen deficit in both STF reactors was mostly due to autotrophic nitrogen removal over nitrite.

5.3.5. Operational parameters and conditions for successful autotrophic nitrogen removal over nitrite in STF systems

5.3.5.1. Temperature

According to Hellinga *et al.* (1998), the growth rate of AOO is higher than that of NOO at higher temperatures, i.e. 30-35°C; thus nitrite accumulation can be induced by taking advantage of temperature. In addition, the appropriate temperature range for Anammox bacteria growth is 20-45°C, which is species dependant (Kartal *et al.*, 2012). For instance, Hendrickx *et al.* (2014) found an optimum temperature for *Candidatus Brocadia fulgida* between 20-30°C. The STF reactors were operated in a temperature controlled room at 30°C which likely enhanced the immobilization and development of AOO and Anammox mainly in the first upper half of both reactors. This is an advantageous feature for the application of STF reactors in warm climates.

Table 5.4. Results of the Anammox activity test executed in the Sponge-bed Trickling Filter 1 during phase II-D.

Reactor	Parameter (unit)	Sample taken after		
		3 h	4 h	5 h
STF-1	Influent NH_4^+ (mg/L)	53.03	50.18	48.71
	Influent NO_2^- (mg/L)	42.81	51.39	51.90
	Total N influent (mg/L)	95.84	101.56	100.60
	NLR (kg N/m^3_{sponge}·d)	1.44	1.53	1.51
	Effluent NH_4^+ (mg/L)	14.45	7.01	5.45
	Effluent NO_2^- (mg/L)	0.10	0.13	0.13
	Effluent NO_3^- (mg/L)	15.21	14.32	13.36
	Total N effluent (mg/L)	29.76	21.47	18.94
	N removal (mg/L)	66.09	80.09	81.66
	N removal (%)	68.95	78.86	81.17
	N removal (kg-N/m^3_{sponge}·d)	0.99	1.21	1.23
	[a]NO_2^--N consumed (mg/L)	42.71	51.25	51.77
	[a]NH_4^+-N consumed (mg/L)	37.27	44.72	45.18
	[a]NO_3^--N produced (mg/L)	6.00	7.20	7.27

[a]Assuming Anammox as dominant microorganism for nitrogen removal and according to the stoichiometry proposed by Lotti *et al.* (2014).

Table 5.5. Results of the Anammox activity test executed in the Sponge-bed Trickling Filter 2 during phase II-D.

Reactor	Parameter (unit)	Sample taken after		
		4 h	**5 h**	**6 h**
	Influent NH_4^+ (mg/L)	74.78	71.19	68.60
	Influent NO_2^- (mg/L)	70.32	73.78	68.66
	Total N influent (mg/L)	145.11	144.97	137.26
	NLR (kg N/m^3_{sponge}·d)	1.16	1.16	1.10
	Effluent NH_4^+ (mg/L)	15.57	7.18	3.25
	Effluent NO_2^- (mg/L)	0.14	0.27	0.20
	Effluent NO_3^- (mg/L)	25.18	20.49	15.25
STF-2	Total N effluent (mg/L)	40.89	27.95	18.69
	N removal (mg/L)	104.21	117.02	118.56
	N removal (%)	71.82	80.72	86.38
	N removal (kg-N/m^3_{sponge}·d)	0.83	0.94	0.95
	[a]NO_2^--N consumed (mg/L)	70.18	73.50	68.46
	[a]NH_4^+-N consumed (mg/L)	61.24	64.14	59.74
	[a]NO_3^- - N produced (mg/L)	11.30	11.83	11.02

[a]Assuming Anammox as dominant microorganism for nitrogen removal and according to the stoichiometry proposed by Lotti *et al.* (2014).

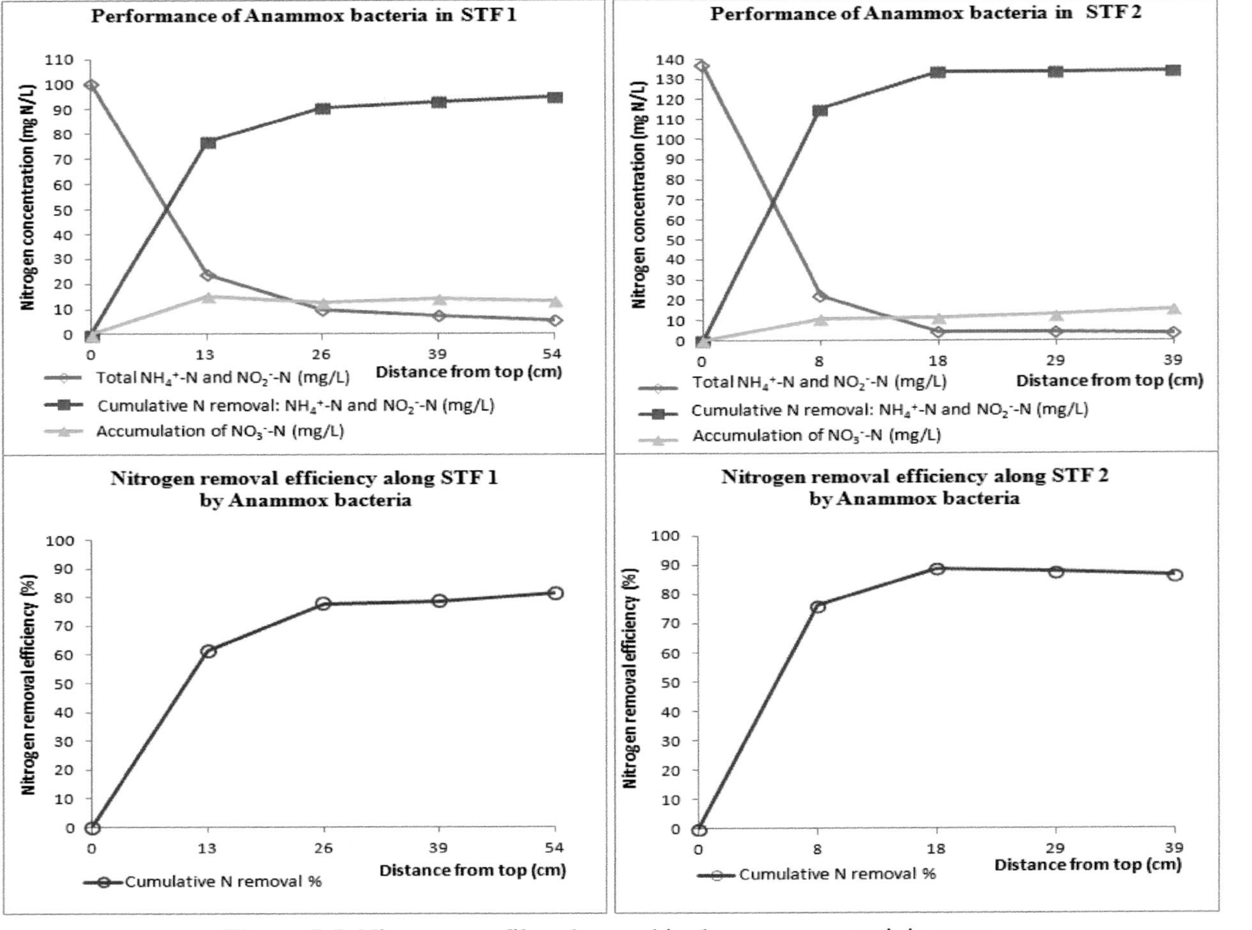

Figure 5.5. Nitrogen profiles observed in the anammox activity tests.

5.3.5.2. Sponge thickness and DO

Uemura *et al.* (2012) found a reciprocal correlation between the sponge media sizes and the NH_4^+-N removal efficiency in Down-flow Hanging Sponge (DHS) reactors, which was attributed to a higher oxygen uptake in small-sized sponges and a better contact between biomass and wastewater. In addition, the DO variation across the sponge media also plays a vital role on nitrogen removal by AOO. Araki *et al.* (1999) showed that sponge cubes (of around 1.5x1.5x1.5 cm) maintain an anoxic environment beyond the depth of 0.75 cm from the surface.

Similarly, Machdar *et al.* (2000) demonstrated that the triangular prism shaped sponge of side 3cm and length 75 cm, maintained aerobic conditions down to a depth of 1.0 cm. However, Machdar *et al.* (2000) and Uemura *et al.* (2012) did not observe significant nitrification when D.O. dropped below 1 mg-O_2/L (in sponge depths of around 50 cm). Moreover, Chuang *et al.* (2007) achieved partial nitrification in a closed DHS reactor that was operated with an oxygen content in the gas phase below 1% resulting in 0.42 mg/L of DO, HRT of 1.5h, temperature of 30°C, and 100 mg-N/L of influent ammonium. In the STF reactors, during most of Phase II the average measured DO in STF-1 was around 1.8-2.3 mg-O_2/L whereas for STF-2 between 2.0-2.2 mg-O_2/L (Figure 5.6). These conditions, in combination with the NLR applied, appeared to be stimulatory to the development of a system where not only nitritation was achieved but also autotrophic nitrogen removal occurred with natural air convection without the associated costs of investment in infrastructure, equipment, maintenance and energy for controlled aeration. Moreover, the observed successful operation of STF at comparatively higher DO levels under natural air convection is an added advantage and indicates flexibility for the STF system at practical applications. Further research must focus on the improvement of the natural aeration process by assessing the oxygen transfer efficiency between the sponge layers, the ammonium concentrations, and their relationship with the oxygen required to achieve partial nitritation.

5.3.5.3. pH and alkalinity

The influent pH was fairly stable at about 7.85 ± 0.12 and 7.96 ± 0.10 for STF-1 and STF-2, respectively. Between influent (L_0) and level L_1 (13 cm from the top), the average pH drop was identical in both reactors (0.56 pH units), and in average in the two consecutive levels (L) a drop of 0.20 pH units was observed (similar for both STF-1 and STF-2).

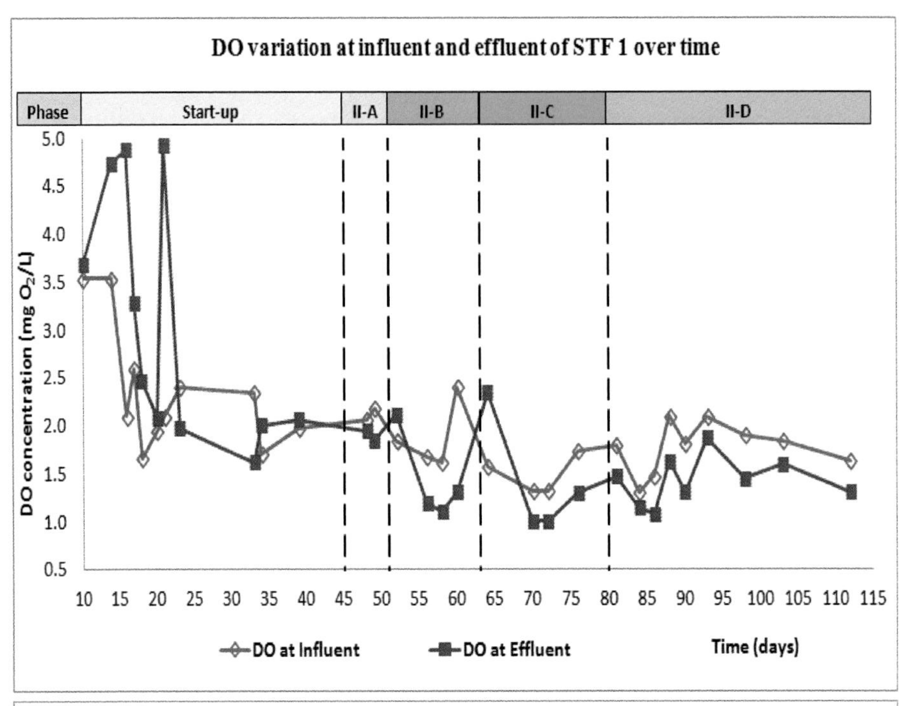

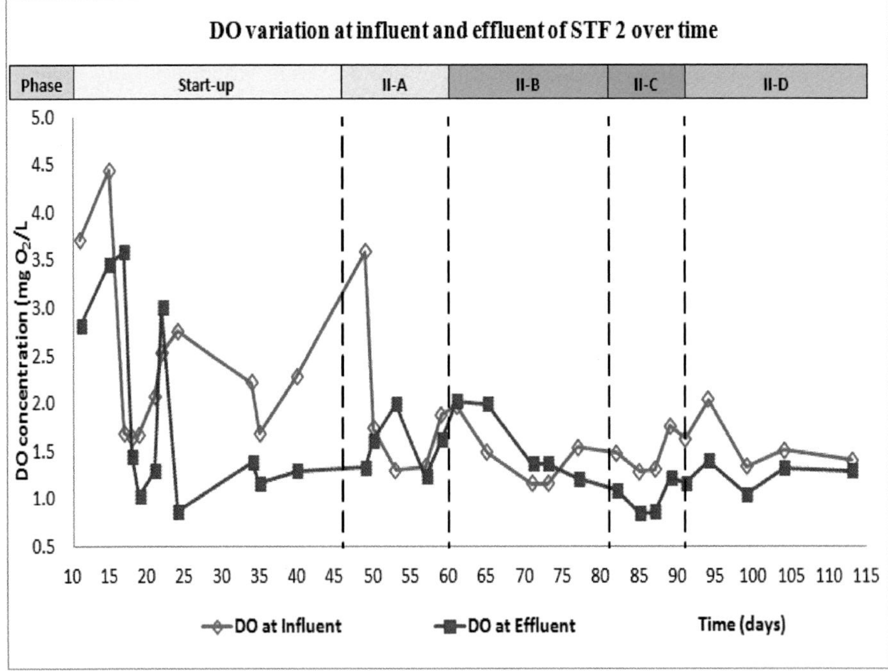

Figure 5.6. DO variation at influent and effluent of STF reactors over time.

These values are indicative for a higher microbiological activity in the first section (L_0-L_1). Finally, the effluent pH reached average values of 6.69 ± 0.20 in STF-1 and 6.86 ± 0.12 in STF-2.

Biological ammonium oxidation is influenced by pH and the optimum range for AOO and NOO lies between 7 to 8 (van Hulle *et al.*, 2010). In addition, Anthonisen *et al.* (1976) observed the inhibition of NOO at free ammonia (FA) concentrations of 0.08-0.82 mg/L, while AOO was inhibited at 8-120 mg/L. Though the influent pH led to an influent FA concentration of about 6-8 mg/L, which according to Anthonisen *et al.* (1976) could have led to NOO inhibition, NOO were active and nitrate production was observed from level L_1 in both STF reactors, e.g. of up to 12 ± 8 mg/L and 10 ± 1 mg/L for STF-1 and STF-2, respectively, during Phase II-D. These results are in accordance with the conclusion of Hawkins *et al.* (2012) that stated that FA does not significantly inhibits NOO; the NOO inhibition is mostly the result of poor competition for oxygen with a very active AOO population.

Nevertheless, aiming at improving the reactor performance and assuming that there are no limitations by neither micronutrients nor alkalinity (alkalinity in the effluent was 215 mg $CaCO_3$/L and 320 mg-$CaCO_3$/L for STF-2 and STF-1, correspondingly), increasing the pH (e.g. to 7.8-8.1) from the level L_2 could lead to a higher performance since the growth rate of ammonia oxidizers is higher than that of nitrite oxidizers at pH levels around 7.9–8.2 (Hellinga *et al.*, 1998). Further research is needed concerning the effects of pH and pH gradients on the microbial populations in sponge-bed systems.

5.3.5.4. NLR and HRT

Okabe *et al.* (2011) observed a gradual increase in nitrite production when increasing the NLR in an up-flow biofilm reactor with nonwoven fabric sheets, with a negligible nitrate production rate at an NLR higher than 1.0 kg-NH_4^+-N/m^3·d. In addition, Chuang *et al* (2007) found that in a DHS reactor nitritation increased with the increase in NLR from 0.47 to 6.42 kg-NH_4^+-N/m^3_{sponge}·d. In both cases, the NLR was raised by significantly increasing the concentration of ammonium in the influent. In the study with the STFs, the NH_4^+-N concentration in the influent was kept constant while the influent flow was increased to follow a realistic approach for the evaluation of the NLR during the operation of the STF reactors. The most remarkable observation in the STFs was the considerable nitrogen removal achieved in both reactors in only 100 days of reactor operation. STF-1 showed an average nitrogen removal capacity of 52.2% (0.88 kg-N/m^3_{sponge}·d) at the NLR of 1.68 kg-N/m^3_{sponge}·d and HRT of 1.71h; whereas, STF-2 had 54% nitrogen removal efficiency (0.51 kg-N/m^3_{sponge}·d) at the NLR of 0.95 kg-N/m^3_{sponge}·d and HRT of 2.96 h. Table 5.6. shows the summary of the nitrogen removal performance by the two reactors. Potentially, a higher NLR could have been applied to further increase the ammonium oxidation to nitrite, followed by a consequently higher Anammox activity.

With reference to the HRT applied in other autotrophic nitrogen removal systems; the STF reactors possess a short HRT with a high nitrogen removal capacity. For example, with an actual $HRT_{sponge\ volume\ based}$ of 1.71 and 2.96 h, the maximum NH_4^+-N removal at phase II-D was 1.22 and 0.83 kg-N/m^3_{sponge}·d in STF-1 (NLR: 1.68 kg-N/m^3_{sponge}·d) and STF-2 (NLR: 0.95 kg-N/m^3_{sponge}·d), respectively (Table 5.6). In contrast, from a DHS system for partial nitritation with an HRT of 1.5 h and NLRs of 1.60 and 0.84 kg-N/m^3_{sponge}·d, Chuang *et al.* (2007) reported a NH_4^+-N removal in the order of 0.77 and 0.42 kg-N/m^3_{sponge}·d, correspondingly. A short HRT of 3.2 h and a long SRT (> 135 days) in a DHS reactor with polyurethane sponge also have been carried out by Onodera *et al.* (2013). Nevertheless, the maximum removal of dissolved inorganic nitrogen (DIN) of this system was only 5 mg N/$L_{influent}$ whereas in the STF reactors 66-71 mg-N/$L_{influent}$ were achieved (Table 5.6).

Additionally, the research of Barana *et al.* (2013) illustrates how Anammox bacteria can grow in systems not deliberately designed for that purpose, as long as the necessary conditions for Anammox growth are provided. This group of researchers used a fixed-bed reactor with cylindrical tubes of polyurethane sponge as biomass support media (2cm diameter and 70 cm length) and mechanical aeration was applied. This reactor was used as post-treatment for the effluent of a UASB reactor utilized for the treatment of poultry slaughterhouse effluents. The fixed-bed reactor achieved a NH_4^+-N removal of 0.163 kg-N/m^3_{sponge}·d at an HRT of 24 h. Compared to this reactor, again the results from STF reactors show a short HRT and a distinctly higher NH_4^+-N removal. The immobilization and growth of the biomass on the polyurethane sponge medium allowed the operation of STF reactors at short HRTs. Therefore, the STF systems only require a very small footprint for construction in addition to its high nitrogen removal capacity.

If only the NH_4^+-N removal efficiency is taken into account, STF-2 had higher removal efficiency than STF-1. But, in terms of total nitrogen removal, both reactors have a similar removal performance although STF-1 presented a higher nitrogen

Table 5.6. Nitrogen removal performance of the Sponge-bed Trickling Filters at phase II-D.

Parameter (unit)	STF-1	STF-2
	Average (maximum)	Average (maximum)
Sponge thickness (cm)	0.75	1.5
NLR (kg-N/m^3_{sponge}·d)	1.68	0.95
Actual $HRT_{sponge\ volume\ based}$ (h)	1.71	2.96
Air circulation	No air supply to 7 sponge sheets	Fully opened
Influent NH_4^+-N (mg-N/L)	111.9 ± 5.5	118.5 ± 4.6
Effluent NH_4^+-N (mg-N/L)	34.3 ± 3.6	21.7 ± 3.6
Effluent NO_2^--N (mg-N/L)	0.3 ± 0.1	0.4 ± 0.1
Effluent NO_3^--N (mg-N/L)	18.9 ± 3.4	32.4 ± 1.1
NH_4^+-N removal (kg-N/m^3_{sponge}·d)	1.17 (1.22)	0.77 (0.83)
NH_4^+-N removal efficiency (%)	69.3 (71.7)	81.6 (86.7)
N removal (mg/L)	58.4 (66)	64.1 (70.9)
N removal (kg-N/m^3_{sponge}·d)	0.88 (0.99)	0.51 (0.57)
N removal efficiency (%)	52.2 (60.4)	54 (61.9)
N removal/ NLR	0.52	0.54
Estimated NO_3^--N produced due to Anammox activity (mg-N/L)	9.4	10.3

removal rate for a given higher NLR with a short HRT. Since the initial objective of the present research was to achieve partial nitritation under natural air convection using different sponge depths, the operational strategies adopted were not suitable to enhance the nitrogen removal up to Phase II-C. However, the operational strategies created in Phase II-D favoured the nitrogen removal in both reactors. It is significant that the ratio N removal/NLR (kg-$N_{removed}$ $m^{-3}{}_{sponge}$ d^{-1}/ kg-$N_{supplied}$ $m^{-3}{}_{sponge}$ d^{-1}) at the end of Phase II-D is the same for both reactors (Table 5.6).

Nevertheless, it is difficult to make a direct comparison of the efficiency of STF-1 and STF-2 because of the different applied NLRs and HRTs and likely differences in air convective flows. For a fair comparison, the performance of both reactors needs to be evaluated under exactly the same operating conditions, e.g. identical NLR and identical airflows. However, the results obtained from the STFs demonstrate that an appropriate convective airflow is the driving force of the partial nitritation done by the AOO present in the STF reactors since the used sponge thickness did not play a crucial role in the nitrogen removal achieved, i.e. the ratio N removal/NLR was the same in STF-1 and STF-2 in Phase II-D. Possibly, in a full scale-DHS reactor the control of the convective air supply would allow to create specific zones favorable for partial ammonium oxidation to nitrite. Thus, by minimizing the oxygen supply, these zones could be extended and gain a major nitrogen removal via Anammox bacteria.

Furthermore, an effective control of the airflow promotes the development of AOO and Anammox bacteria in the sponge in such way that a straightforward sequence of AOO and Anammox bacteria in each sponge layer is apparently not required. This additional feature simplifies the application of the STF reactor. Additionally, the total NO_3^--N concentration in the effluents is 2 and 3 times higher than the NO_3^--N produced by Anammox activity in STF-1 and STF-2, respectively (Table 5.6). This is indicative of a high nitrification in STF-2, which is also confirmed by the recorded alkalinity consumption. On the other hand, the fact that the lower 1/3rd of the STF-1 height did not have additional air inlet points might have caused the lower nitrification activity observed.

The gradual increase in NO_2^--N observed with respect to the low applied NLR (0.65 kg-N/ $m^3{}_{sponge}$·d) during the start-up phase in both STF systems, contributed to the development of the nitrogen removal capacity after the short start-up period. This result is in agreement with the statement of van Hulle *et al.* (2011) who concluded that for achieving good Anammox activity a slow feeding regime is necessary in the start-up period. Therefore, an NLR of about 0.65 kg-N/ $m^3{}_{sponge}$·d could be suggested as the lower limit of NLR for the start-up in STF systems with the aim to achieve a single stage autotrophic nitrogen removal over nitrite. For further assessing the optimum and upper limit of NLRs, meanwhile achieving successful nitrogen removal, extension of the current research in the STFs is necessary.

5.3.6. Future perspective of SFT bioreactors

The group of Harada and collaborators already proposed in 1997 for the first time a sponge-based bioreactor, denominated as Down-flow Hanging Sponge (DHS), as a novel cost-effective post-treatment method for anaerobically pre-treated sewage (Machdar *et al.*, 1997). Since then multiple investigations have shown the effectiveness of a UASB-DHS combined system during sewage treatment at lab, pilot-scale and large-scale (Machdar *et al.*, 2000; Harada, 2014; Khan *et al.*, 2014), reaching removal efficiencies of 70% for TSS, 80% for total COD and 95% for total biochemical oxygen demand (BOD) (Tandukar *et al.*, 2006). Regarding to the DHS reactor performance, the microbial community composition has revealed that organic matter removal significantly occurs in the upper part of DHS, while the NH_4^+-N conversion is mainly accomplished in the lower sections via nitrification and denitrification (Kubota *et al.*, 2014). The extent of nitrogen removal by denitrification is variable and depends on the features of each DHS reactor and the operational regimes.

The DIN removal achieved in DHS reactors when applied as post-treatment of anaerobic sewage treatment by UASB reactors, have been accounted for about 22% (Onodera *et al.*, 2013 and 2014) and 44% (Tandukar *et al.*, 2006). The removal efficiency of DIN in sewage treatment by the UASB-DHS system might comply with the standards for effluent discharge, e.g. in some developing countries (Mahmoud *et al.*, 2011). However, in circumstances where the implementation of regulations is more stringent, for instance when a concentration of total nitrogen in the effluent of less than 10-15 mg/L is required (Oleszkiewicz and Barnard, 2006), the DIN in the effluent from a UASB-DHS system might be higher than this environmental regulation, e.g. 21 mg-DIN/L (Tandukar *et al.*, 2006; Onodera *et al.*, 2013) and 16 mg-DIN/L (Onodera *et al.*, 2014). In the conventionally designed DHS, the limitation of available COD could be the reason for the remaining NO_3^--N in the effluent, i.e. a deficit of COD for denitrification in the order of 10% and 77% was observed by Tandukar *et al.* (2006) and Onodera *et al.* (2013), respectively. In addition, the NH_4^+-N is also present in the DHS's effluent and may be an important part of the DIN, i.e. 48% (Tandukar *et al.*, 2006) and 73% (Onodera *et al.*, 2014) in the latter researches.

Given the low COD content and the presence of NH_4^+-N in the DHS's effluent, the STF reactor might be a suitable DHS amendment for the treatment of UASB effluents, introducing autotrophic nitrogen removal over nitrite in the sponge-bed. Such amendment could possibly be achieved by adding additional layers to the height of the sponge-bed filter, or by placing filters in series. In the proposed scenario, the nitrifying microorganisms would not be suppressed by heterotrophs because of the low COD as long as enough natural air convection is guaranteed and satisfies the requirements of both microbial populations. Furthermore, based on the results obtained in the STF reactors, it is expected that autotrophic NH_4^+-N removal will rapidly develop without the need for inoculation with Anammox enriched sludge and/or long start-up periods. Moreover, the proposed system is operated without mechanical operation or expensive aeration control systems, whereas a high nitrogen removal at a small footprint is achieved.

5.4. Conclusions

A single stage autotrophic nitrogen removal over nitrite using STF reactors under natural air convection is technically feasible. Nitrogen removal (52-54%) was observed resulting from the coexistence of AOO and Anammox bacteria and was governed by the DO (1.5-2 mg-O_2/L) and pH (about 8.0). Both STF reactors showed robustness to NLR fluctuations, short HRT (1.71-2.96 h) and had a similar nitrogen removal capacity. The STF or "extended DHS" bioreactor is a promising technology that could be coupled to a UASB reactor to develop a cost-effective post-treatment system for ammonium removal provided extensive organic matter removal is achieved upstream.

References

Anthonisen A.C., Loehr R.C., Prakasam T.B.S., Srinath E.G., 1976. Inhibition of nitrification by ammonia and nitrous acid. J. Water Pollut. Control Federation, 48 (5), 835-852.

Araki N., Ohashi A., Machdar I., Harada H., 1999. Behaviors of nitrifiers in a novel biofilm reactor employing hanging sponge-cubes as attachment site. Water Science and Technology, 39 (7), 23-31.

Barana A.C., Lopes D.D., Martins T.H., Pozzi E., Damianovic M.H.R.Z., Del Nery V., Foresti E., 2013. Nitrogen and organic matter removal in an intermittently aerated fixed-bed reactor for post-treatment of anaerobic effluent from a slaughterhouse wastewater treatment plant. J. Environ. Chem. Eng., 1, 453-459.

Chernicharo C.A.L., Almeida P.G.S., Lobato L.C.S., Rosa A.P., 2012. Anaerobic domestic wastewater treatment in Brazil: drawbacks, advances and perspectives. Water 21, October 2012, 24-26.

Chernicharo C.A.L., van Lier J.B., Noyola A., Bressani Ribeiro T., 2015. Anaerobic sewage treatment: state of the art, constraints and challenges. Reviews in Environmental Science and Bio/technology, 14 (4), 649-679.

Chuang H.P., Ohashi A., Imachi H., Tandukar M., Harada H., 2007. Effective partial nitrification to nitrite by down-flow hanging sponge reactor under limited oxygen condition. Water Research, 41, 295-302.

Ekama G. A., Wentzel M.C., 2008. Nitrogen Removal. In: Biological Wastewater Treatment: Principles, Modelling and Design, M. Henze, M.C.M. van Loosdrecht, G.A. Ekama and D. Brdjanovic (eds.), IWA Publishing, London, UK, pp. 87-138.

Harada H., 2014. UASB-DHS integrated system: a sustainable sewage treatment technology. In: Proc. of International Workshop on "UASB-DHS integrated system – a Sustainable Sewage Treatment Technology", 16-18 October, 2014, Agra, New Delhi & Agra, India.

Hawkins S., Robinson K., Layton A., Sayler G., 2012. Molecular indicators of *Nitrobacter* spp. population and growth activity during an induced inhibition event in a bench scale nitrification reactor. Water Research, 46, 1793-1802.

Hellinga C., Schellen A.A.J.C., Mulder J.W., Van Loosdrecht M.C.M., Heijnen J.J., 1998. The SHARON process: an innovative method for nitrogen removal from ammonium-rich waste water. Water Science and Technology, 37, 135-142.

Hendrickx T.L.G., Kampman C., Zeeman G., Temmink H., Hu Z., Kartal B., Buisman C.J.N., 2014. High specific activity for anammox bacteria enriched from activated sludge at 10^0C. Bioresource Technology, 163, 214-221.

ISO 7890/1-1986 (E). Water quality - Determination of nitrate - Part 1: 2, 6-Dimethylphenol spectrometric method.

Kartal B., van Niftrik L., Keltjens J.T., Op den Camp H.J.M., Jetten S.M., 2012. Anammox-Growth physiology, cell biology, and metabolism. In: Advances in

Microbial Physiology, R.K. Poole (ed.), vol. 60. University of Sheffield, UK., pp. 211-262.

Khan A.A., Gaur R.Z., Mehrotra I., Diamantis V., Lew B., Kazmi A.A., 2014. Performance assessment of different STPs based on UASB followed by aerobic post treatment systems. J. Environ. Health Sci. Eng., 12 (43), 1-13.

Kubota K., Hayashi M., Matsunaga K., Iguchi A., Ohashi A., Li Y.Y., Yamaguchi T., Harada H., 2014. Microbial community composition of a down-flow hanging sponge (DHS) reactor combined with an up-flow anaerobic sludge blanket (UASB) reactor for the treatment of municipal sewage. Bioresource Technology, 151, 144-150.

Lackner S., Gilbert E.M., Vlaeminck S.E., Joss A., Horn H., van Loosdrecht M.C.M., 2014. Full-scale partial nitritation/anammox experiences - An application survey. Water Research, 55, 292-303.

Levenspiel O., 1999. Chemical Reaction Engineering. John Wiley & Sons, New York. USA, pp. 257-282.

Lotti T., Kleerebezem R., Lubello C., van Loosdrecht M.C.M., 2014. Physiological and kinetic characterization of a suspended cell anammox culture. Water Research, 60, 1 -14.

Machdar I., Harada H., Ohashi A., Sekiguchi Y., Okui H., Ueki K., 1997. A novel and cost-effective sewage treatment system consisting of UASB pre-treatment and aerobic post-treatment units for developing countries. Water Science and Technology, 36 (12), 189-197.

Machdar I., Sekiguchi Y., Sumino H., Ohashi A., Harada H., 2000. Combination of a UASB reactor and a curtain type DHS (downflow hanging sponge) reactor as a cost-effective sewage treatment system for developing countries. Water Science and Technology, 42 (3-4), 83-88.

Mahmoud M., Tawfik A., El-Gohary F., 2011. Use of down-flow hanging sponge (DHS) reactor as a promising post-treatment system for municipal wastewater. Chem. Eng. J., 168, 535-543.

NEN, 1983. Photometric determination of ammonia in Dutch system. In: Nederlandse Normen (Dutch Standards), International Organization for Standardization (ed.) NEN 6472, Dutch Institute of Normalization, Delft, the Netherlands.

Okabe S., Oshiki M., Takahashi Y., Satoh H., 2011. Development of long-term stable partial nitrification and subsequent anammox process. Bioresource Technology, 102, 6801-6807.

Oleszkiewicz J.A., Barnard J.L., 2006. Nutrient Removal Technology in North America and the European Union: A Review. Water Qual. Res. J. Canada, 41 (4), 449-462.

Onodera T., Matsunaga K., Kubota K., Taniguchi R., Harada H., Syutsubo K., Okubo T., Uemura S., Araki N., Yamada M., Yamauchi M., Yamaguchi T., 2013. Characterization of the retained sludge in a down-flow hanging sponge (DHS) reactor with emphasis on its low excess sludge production. Bioresource Technology, 136, 169-175.

Onodera T., Tandukar M., Sugiyana D., Uemura S., Ohashi A., Harada H., 2014. Development of a sixth-generation down-flow hanging sponge (DHS) reactor using rigid sponge media for post-treatment of UASB treating municipal sewage. Bioresource Technology, 152, 93-100.

Schmid M., Twachtmann U., Klein M., Strous M., Juretschko S., Jetten M., Metzger J.W., Schleifer K.H., Wagner M., 2000. Molecular evidence for genus level diversity of bacteria capable of catalyzing anaerobic ammonium oxidation. Syst. Appl. Microbiol., 23 (1), 93-106.

Standard Methods for the Examination of Water and Wastewater 2012 22th edn, American Public Health Association/American Water Works Association/Water Environment Federation, Washington DC, USA.

Tandukar M., Machdar I., Uemura S., Ohashi A., Harada H., 2006. Potential of a combination of UASB and DHS reactor as a novel sewage treatment system for developing countries: long-term evaluation. J. Environ. Eng., 132, 166-172.

Uemura S., Suzuki S., Abe K., Ohashi A., Harada H., Ito M., Imachi H., Tokutorni T., 2011. Partial nitrification in an airlift activated sludge reactor with experimental and theoretical assessments of the pH gradient inside the sponge support medium. Int. J. Environ. Res., 5 (1), 33-40.

Uemura S., Suzuki S., Maruyama Y., Harada H., 2012. Direct treatment of settled sewage by DHS reactors with different sizes of sponge support media. Int. J. Environ. Res., 6 (1), 25-32.

van de Graaf A.A., de Bruijn P., Robertson L.A., Jetten M.S.M., Kuenen J.G., 1996. Autotrophic growth of anaerobic ammonium-oxidizing micro-organisms in a fluidized bed reactor. Microbiology, 142, 2187-2196.

van Hulle S.W.H., Vandeweyer H.J.P., Boudewijn D., Meesschaert D., Vanrolleghem P.A., Dejans P., Dumoulin A., 2010. Engineering aspects and practical application of autotrophic nitrogen removal from nitrogen rich streams. Chem. Eng. J., 162, 1-20.

van Hulle S.W.H., Vandeweyer H., Audenaert W., Monballiu A., Meesschaert B., 2011. Influence of the feeding regime on the start-up and operation of the autotrophic nitrogen removal process. Water SA, 37, 289-294.

van Lier J.B., Vashi A., van der Lubbe J., Heffernan B., 2010. Anaerobic sewage treatment using UASB reactors: engineering and operational aspects. In Environmental Anaerobic technology; Applications and New Developments (H.H.P. Fang ed.), World Scientific, Imperial College Press, London, UK, pp. 59-89.

Wilsenach J., Burke L., Radebe V., Mashego M., Stone W., Mouton M., Botha A., 2014. Anaerobic ammonium oxidation in the old trickling filters at Daspoort Wastewater Treatment Works. Water SA, 40 (1), 81-88.

Evaluation and Outlook

Evaluation and Outlook

Diverse challenges need to be faced for achieving DIN removal, via Anammox-based technology, during the post-treatment of the effluents from UASB reactors used for treating sewage. This research has chosen some of these challenges in terms of operational features and system configuration. Regarding these aspects, the water quality of the UASB effluent plays an important role because it sets the boundary conditions for the functionality of the post-treatment system to be designed. In UASB reactors organic matter is stabilized and, if biodegradable, mineralized leaving residual BOD, TSS, sulfide and inorganic nutrients in the effluent (Chernicharo *et al.*, 2015). Overall BOD, COD and TSS removal efficiencies of UASB reactors treating municipal sewage are in the range of 70-80%, 65-75% and 65-75%, respectively (van Lier *et al.*, 2010; Chernicharo *et al.*, 2015).

The exact treatment efficiencies and therewith, effluent composition, differs from location to location owing to a number of factors (Chernicharo *et al.*, 2015). UASB effluent characterization is the first step to define the strategies for reaching the objectives of the post-treatment, which is commonly removal of residual BOD, TSS and the nutrients. Particularly DIN removal is considered a major challenge since conventional nitrogen removal, following the nitrification-denitrification approach, is simply not possible after UASB pre-treatment. However, the recent developments in ***autotrophic*** DIN removal give ample opportunities for combing cost-effective anaerobic pre-treatment with efficient DIN removal techniques. In order to evaluate the possible impact of UASB effluent composition on autotrophic DIN removal systems, a complete or partial characterization of UASB effluents in full-scale or pilot-plant treatment facilities is required. However, such characterization is not or rarely found in the literature, unless the research objectives were targeting specific constituents in the UASB effluent. For instance, Procópio Pontes and Chernicharo (2011) reported the concentrations of carbohydrates, proteins and lipids in the UASB effluent that was treated by a trickling filter (demo-scale system for 500 inhabitants) with the aim of evaluating the influence of the sludge recirculation in the removal of these compounds.

The major objective of this research was to evaluate the potential of partial nitritation and autotrophic DIN removal from anaerobic effluents that are characterized by relatively low concentrations of ammonia (up to 100 mg/L). In order to reduce the degrees of freedom during this research, synthetic mineral media without organic matter was used to simulate the UASB effluent. In any follow up research, the use of more complex substrate is recommended to study in more detail the long-term performance and stability of the envisaged post-treatment system. Nonetheless, in the present thesis research several effluent characteristics were taken into account. **Chapter 2** describes the simultaneous effects of the organic matter composition, temperature and COD/N ratios on autotrophic nitrogen removal. In the related experiments the synthetic media was complemented with different organic compounds simulating either readily or slowly biodegradable matter, i.e. acetate as readily biodegradable COD (RBCOD) and starch as slowly biodegradable COD (SBCOD).

Results showed that none of these COD fractions affected the metabolism of the Anammox consortia in the batch tests, which were performed at different COD/N ratios and temperatures. Similarly, the presence of denitrifying organisms did not impact the performance of Anammox biomass. However, it must be noted that long-term studies are necessary to elucidate the degree of competition between the microbial populations involved in the metabolic pathways of DIN removal under the influence of the different COD fractions.

Likely, the characterization and quantification of the dominant COD fractions, i.e. RBCOD and SBCOD, from a full-scale UASB effluent would provide accurate and sensitive information to be applied in such long-term experiments. In addition, variations in both composition and ratios of the RBCOD and SBCOD fractions, while analyzing the response of the microorganisms involved to each new scenario, will provide useful information regarding long-term reactor stability. Under dynamic conditions, even changes in the trophic bacterial subpopulation responsible for DIN removal are expected. For example, different species of Anammox bacteria are able to metabolize organic compounds such as acetate, propionate and formate (**Chapters 1 and 2**), whereas the survival of some Anammox bacteria are based on the maximization of their activity or their substrate affinity (Kartal *et al.*, 2012).

With reference to these survival strategies, van der Star *et al.* (2008) found in a membrane bioreactor (MBR) with Anammox bacteria living as suspended free cells, nitrogen influent concentrations of 1680 mg-NO_2^--N/L and 1400-1680 mg-NH_4^+-N/L and an HRT of 2 days that a shift in the Anammox population occurred from *Candidatus Brocadia anammoxidans* to *Candidatus Kuenenia stuttgartiensis*. This outcome was explained by the fact that *Candidatus Kuenenia stuttgartiensis* has a lower nitrite apparent half-saturation constant (K_m: 0.2-3 μM) than *Candidatus Brocadia anammoxidans* (K_m: <5 μM). Also Kartal *et al.* (2012) obtained the same result. In terms of the dominant Anammox population developing in the SBR, while applying an ultra low nitrogen sludge loading rate (NSLR) (**Chapter 3**), we expected the dominance of a species with the lowest possible half saturation constant. However, despite the fact that the Anammox SBR was operated under nitrite and ammonium limitation applying a long SRT, there was no population shift in the Anammox species, i.e. the *Candidatus Brocadia fulgida* remained as the dominant species for more than 1000 days. Lotti *et al.* (2014) reported a nitrite half-saturation constant of 2.5 μM in a free cells culture with an enriched level of 98 ± 1% for *Candidatus Brocadia fulgida.* This value for the nitrite half-saturation constant is lower than other values reported for Anammox species (Puyol *et al.*, 2013), but it might be not low enough compared to the range reported for *Candidatus Kuenenia stuttgartiensis* (0.2-3 μM). Additional research concerning the dominance of *Candidatus Brocadia fulgida* over other Anammox species at very low NSLR is recommended.

Anammox bacteria are slow growing organisms that have a low biomass yield. Taking into account this feature, the effective retention of Anammox biomass inside the wastewater treatment facilities is one of the main challenges for the application of Anammox technology, particularly in main-stream treatment systems. Sponge-based reactors have been investigated for more than 15 years and the first full-scale reactor for sewage treatment is in operation in India since 2014. The closed sponge trickling filter (CSTF; **Chapter 4**) was investigated as a first step for the immobilization and cultivation of Anammox bacteria in sponge-bed reactors with the aim to offer a low-cost alternative for DIN removal from sewage.

The impact of substrate concentrations on the performance of the CSTF reactors was considered to be evaluated by the four consecutive compartments that compose each reactor. In fact, these compartments represent four CSTF in series, with 95 sponge cubes each one, at the same temperature and operational conditions. The long term operation, which was more than 400 days, demonstrated consistent results for both reactors at each temperature, i.e. 20°C and 30°C. Therefore, we considered the experimental design and execution of the research enough robust to draw the conclusions obtained in **Chapter 4** of this thesis.

However, there are several operational conditions and design aspects that need to be investigated further to provide a more complete picture of the CSTF. For instance, to what extent the recirculation rate is necessary? Is it possible to apply the recirculation of the effluent only during certain periods, e.g. only during the start-up phase? In the same way, taking into account that in the downflow hanging sponge (DHS) reactors, the excess sludge production rate is very low compared to other treatment systems as activated sludge; 0.09 *vs.* 0.88 g-TSS/g-$COD_{removed}$, respectively (Onodera *et al.*, 2013), a long-term study of the sludge production in the CSTF will supply useful insights of the system. Furthermore, in the present design of the CSTF reactors, the applied sponge volume occupancy had to satisfy two practical criteria: i) to allow easy access to each section of the reactor for solving operational problems, for instance the control of clogging by means of the mechanical detachment of the biomass, ii) to avoid compression of the sponge's cubes due to the weight caused by a big number of sponge cubes inside each section. On the other hand, the amount of sponges contained in the CSTF directly relates to an optimum nitrogen loading rate (NLR) applied.

In this research (**Chapter 4**), the CSTF systems almost reached the maximum removal efficiency related to the applied NLR. Results indicate that an increment in the number of sponge cubes could have increased the removal efficiency. Alternatively, a bigger number of cubes will allow treating a higher NLR. Therefore, assessment of the specific attainable DIN removal rate per amount of support material (sponge cubes) needs to be researched under field conditions. Additionally, a detailed study is required on the spatial distribution of the microbial population during the operation of the CSTF under varying loads of organic matter and dissolved gases, e.g. CH_4 and H_2S, coming from the preceding anaerobic treatment system. Results will help to understand the dynamics and capacity of the system in view of a future full-scale implementation.

The effective retention of Ammonium Oxidizing Organisms (AOOs) in the bioreactors is considered essential for achieving a successful partial nitritation that would allow the implementation of an Anammox conversion process. This research work has demonstrated that Closed Sponge-bed Trickling Filters (CSTFs) are able to efficiently retain the required biomass and sustained Anammox cultivation (**Chapter 4**). The next step of the research was to assess the feasibility to attain partial nitritation with natural air convection in lab-scale sponge-bed trickling filters (STFs). During the experiments, a considerable nitrogen removal was observed and experiments were redirected to optimise autotrophic nitrogen removal over nitrite in the STFs (**Chapter 5**).

The biological reactors, at lab or full-scale, require the inoculation of the biomass and a period for the adaptation of the inoculum before achieving the maximum metabolic activity. This procedure can be carried out using biomass from a similar reactor, or inoculation can be done with a type of biomass that would allow the growth and acclimatization of the microorganisms of interest; in our case Anammox and AOOs. One of the outputs from the research with the STFs is the development of AOOs and Anammox from activated sludge under the operational conditions described in

Chapter 5, i.e. there was no necessity of using enriched sludge with AOOs and Anammox bacteria. The obtained results revealed that nitrification was occurring since the beginning of the experiments and nitrogen removal began in a period of 23-36 days; this means that AOOs and Anammox bacteria were active from the early start of the experiment, respectively.

The research in **Chapter 5** was designed for obtaining partial nitritation, i.e. 50% of NH_4^+-N and 50% of NO_2^--N in the effluent. However, after 100 days of operation, including the period for growth and acclimatization of AOOs and Anammox bacteria, approximately 50% of NH_4^+-N was in the effluent and 50% of NH_4^+-N was removed, i.e. the partial nitritation was achieved under natural air convection, but the NO_2^--N produced by the AOOs and part of the NH_4^+-N were removed by Anammox bacteria. Unexpectedly, results demonstrated for the first time the capacity of sponge-bed bioreactors of achieving ammonium removal via ammonium oxidizing organisms (AOO) and Anammox bacteria without mechanical air supply. Consequently, as soon as the consortium of AOOs and Anammox bacteria are established in the STFs, the nitrogen removal can be achieved without any delay.

The sponge bed trickling filter (STF) (**Chapter 5**) is a promising alternative for nitrogen removal using sponge-based reactor technology and a further research for exploring all the capabilities of SFT system for maximizing the level of nitrogen removal is suggested. This novel approach represents an enhancement of the CSTF, since the autotrophic DIN removal can be attained in a single sponge-based reactor that possess a compact configuration. Some of the additional researches that may be done with the STF are analogous to the suggestions previously described for the CSTF reactors, but a crucial investigation is the supply and control of oxygen through natural air convection. From the operational perspective, the understanding of the mechanisms that govern the oxygen mass transfer via natural air convection in the STF is considered critical. Although not an easy task, it is recommended to model the convective airflow in relation to the metabolic activity in the sponge cubes of the STF. Some parameters that likely relate to such modeling approach include air temperature, sewage temperature, geometry of the reactor, number and location of openings for air convection, air flow pattern inside the STF, influent flow, thickness of the sponges, biomass retention, N(S)LR, metabolic conversion rates, etc.

Recently, Scherson *et al.* (2013) proposed a process called the coupled aerobic-anoxic nitrous decomposition operation (CANDO). The process generates N_2O with the aim of using it as a renewable source of energy via two alternatives. One of the alternatives is the decomposition of N_2O by a metal oxide catalyst; the other is the utilization of N_2O as co-oxidant in CH_4 combustion. This combustion produces approximately 30% more heat than the combustion of CH_4 with oxygen and N_2O is reduced to N_2. The UASB-Anammox configuration might explore the feasibility of coupling the combustion of the CH_4 produced by the UASB reactor with the N_2O produced by partial nitritation-Anammox process in order to mitigate the N_2O release to the environment and enhance the energy production capacity of the UASB-Anammox system.

References

Chernicharo C.A.L., van Lier J.B., Noyola A., Bressani Ribeiro T., 2015. Anaerobic sewage treatment: state of the art, constraints and challenges. Reviews in Environmental Science and Bio/technology, 14 (4), 649-679.

Kartal B., van Niftrik L., Keltjens J.T., Op den Camp H.J.M., Jetten M.S.M., 2012. Anammox-growth physiology, cell biology and metabolism. In: Advances in Microbial Physiology (Poole R.K., ed.), Elsevier Ltd., Vol. 60, pp. 211-262.

Lotti T., Kleerebezem R., Lubello C., van Loosdrecht M.C.M., 2014. Physiological and kinetic characterization of a suspended cell anammox culture. Water Research, 60, 1 -14.

Onodera T., Matsunaga K., Kubota K., Taniguchi R., Harada H., Syutsubo K., Okubo T., Uemura S., Araki N., Yamada M., Yamauchi M., Yamaguchi T., 2013. Characterization of the retained sludge in a down-flow hanging sponge (DHS) reactor with emphasis on its low excess sludge production. Bioresource Technology, 136, 169-175.

Procópio Pontes P. and Chernicharo C.A.L., 2011. Characterization and removal of specific organic constituents in an UASB-trickling-filter system treating domestic wastewater. Environmental Technology, 32 (3), 281-287.

Puyol D., Carvajal-Arroyo J.M., García B., Sierra-Alvarez R., Field J.A., 2013. Kinetic characterization of *Brocadia* spp.-dominated anammox cultures. Bioresource Technology, 139, 94-100.

Scherson Y.D., Wells G.F., Woo S-G., Lee J., Park J., Cantwell B.J., Criddle C.S., 2013. Nitrogen removal with energy recovery through N_2O decomposition. Energy Environ. Sci. 6, 241-248.

van der Star, W.R., Miclea A.I., van Dongen U.G., Muyzer G., Picioreanu C., van Loosdrecht M.C., 2008. The membrane bioreactor: a novel tool to grow anammox bacteria as free cells. Biotechnol. Bioeng., 101, 286-294.

van Lier J.B., Vashi A., van der Lubbe J., Heffernan B., 2010. Anaerobic sewage treatment using UASB reactors: engineering and operational aspects. In Environmental Anaerobic technology; Applications and New Developments (H.H.P. Fang ed.), World Scientific, Imperial College Press, London, UK, pp. 59-89.

Acknowledgements

Finishing my PhD. studies has been a long journey and that is why I would start giving thanks to God for all the opportunities given to my academic life.

I would like to express special appreciation to Prof. Jules van Lier for encouraging my research, his continuous support on my study, for his guidance and insightful comments; thank you so much. I would like to thank as well Prof. Damir Brdjanovic and Assoc. Prof. Carlos Lopez Vazquez for sharing their experiences and giving support to my study.

My profound gratitude goes to my family, for their sincere and unconditional support specially my wife and daughter. They are the most important people on my world. I would specially like to thank my dear friend Laurens Welles and his family for their support and always faith in my lab work and for his undoubtedly support. I would thank also my fellow lab-office mates for the stimulating discussions, for the sleepless moments that were working side by side before deadlines and for all the fun! My wife and I have had, thank you! The Sanitary Dreams team: Fiona, Peter, Joy, Yuli, Sondos, Sangyeob and our honorary member Nikola, as well as, all those who we met at UNESCO-IHE. Special thanks to my friend Guy Beaujot for giving the Belgian touch to my stay in the UNESCO-IHE.

Thanks to the laboratory staff, Don, Berend, Ferdi, Peter, Lyzette and Fred. Also my gratitude to my Msc. students, Yesuf Ali Yimmam (Ethiopia), Patricia Cuéllar (El Salvador), Godfrey Baguma (Uganda) and Lansakara K.M.C.B. Jayawardana (Sri Lanka) for their efforts, contributions and friendship.

I would like to express my special appreciation to: Mr. Gerrit Schmidt and his wife who became, for my wife and I, our Dutch mother and father; Mrs. Ria Arkesteyn and Mr. Arnold Arkesteyn (RIP); Theda Olsder and Veronique van der Vaal of TU Delft for their unconditional help. I would not forget Father Visser and his family for his sincere support and for including us in his family activities; always kind and spiritual his advice gave us a glance of light and peace.

I am also hugely thankful to Leen van Ginkel and Linda Stolker; words cannot express and are not enough to say how grateful I am to our Dutch brother and sister. In addition special thanks to Ruud who indirectly always kindly contributed to my work, one more time "thank you so much" for your always kind support. I would like to extend thanks to RIKILT people from Wageningen University, who sometimes without knowing it indirectly so generously contributed to my work by giving support to my wife. Special mention to: Ab, Martien, Hennie, Paula, Efraim, Patrick, Bert, Ivett, Stephan, Els, Coen, Thijs, Michel, Greet, John, Constant, Elly Thissen, Antoine and his wife, Richetti and many others that are not listed here only because it would be too long, thank you so much, thanks to this people.

Thanks for Diego, Alexio, and his people!!!...special food...

Special thanks to Itzel Hubbard and Natacha Gómez who, in Panama, make the difference in the most critical moments of my PhD journey. Also thanks to Catalina and Gilberto for their absolute support.

Thanks to my beloved wife who spent sleepless nights with and was always my support in the moments when there was no one to answer my queries. Last but not least I would like to thank our family, Parents, brothers and sisters, nephews, for their support and good will when writing this thesis. For all the above mentioned it would have never be possible. Thank you so much people!!!!...

Javier.

Agradecer no es fácil porque nunca serán suficientes las palabras para decir todo lo que el corazón desea expresar. Mi agradecimiento por la generosidad, solidaridad y misericordia de quienes nos ayudaron durante esta jornada; sus palabras y actos merecen y necesitan ser reconocidos, valorados y atesorados. Gracias a mis padres Abdiel y Aurelia por todo su amor, acompañamiento y ayuda. Gracias a Efrain e Isaura, los padres generosos que Dios me dio en mi matrimonio, quienes durante mis estudios de doctorado, cuidaron y apoyaron a mi hija María Bernadette. A ti María Bernadette, hija, gracias por creer y esperar. A ti Zahira, esposa mía, todo mi amor y gratitud.

La participación de todos los que nos acompañaron, animaron, ayudaron y guiaron fue decisiva para la culminación de este proyecto. Esta obra no es sólo mía, es también de todos ustedes: mi familia, nuestros familiares, amigos, colegas y supervisores. ¡Gracias!

Javier.

About the Author

Javier Adrián Sánchez Guillén was born in 1964 in La Chorrera, Panama. He got his Chemical Engineering degree in 1988 from the Universidad Autónoma de Puebla, Mexico. After he finished his studies in Mexico, he went back to his country and started working at the National Bureau of Drinking Water Supply and Waste Water Treatment of Panama (IDAAN). As far as work goes, he decided to expand his knowledge in the water sector and traveled to the Netherlands to obtain his Master in Sanitary Engineering at IHE in 1997.

In addition to his education, in 2002 he was awarded with Fulbright fellowship Hubert Humphrey program in Water Quality-Watershed Management which he covered at Cornell University in the United States of America. Also he has been awarded with a scholarship from the Japan International Cooperation Agency (JICA) and the International Lake Environment Committee Foundation (ILEC) in 2004 to study Environmental Education Focused on Fresh Water Environment which he covered at Shiga University in Japan.

Besides his experience as a Chemical Engineer, he also has had the opportunity to work at the Technological University of Panama (UTP), where he has been appointed as a Lecturer in the Master of Science in Environmental Engineering, from 1998 until 2007. During this period he taught the subjects drinking water treatment and wastewater treatment. Later, he decided to go back to the Netherlands to follow his Ph.D. studies on the topic of autotrophic nitrogen removal from low concentrated effluents.

Mr. Sánchez Guillén has a strong interest and a fervent desire to continue with his research in Anammox-based technology with the purpose of contributing with the development of bioprocesses for nitrogen pollution control. After about 22 years of experience in water treatment, he is currently working at the National Bureau of Drinking Water supply and Waste Water Treatment of Panama.

List of Publications

Conference proceedings

Sánchez Guillén J.A., Yimman Y., Lopez Vazquez C.M., Brdjanovic D., van Lier J.B., 2013. Effects of Organic Carbon Source, COD/N ratio and Temperature on Anammox Organisms. In *13th World Congress on Anaerobic Digestion: Recovering (bio) Resources for the World. International Water Association (IWA). June 25-28, 2013. Santiago de Compostela, Spain.*

Sánchez Guillén J.A., Yimman Y., Lopez Vazquez C.M., Brdjanovic D., van Lier J.B., 2013. Effects of Organic Carbon Source, COD/N ratio and Temperature on Anammox Organisms. In *WEF (Water Environment Federation)/IWA Conference in Nutrient Removal and Recovery 2013: Trends in Resource Recovery and Use. July 28-31, 2013. Vancouver, British Columbia. Canada.*

Sánchez Guillén J.A., Cuéllar Guardado P.R., Lopez Vazquez C.M., Brdjanovic D., van Lier J.B., 2014. Low cost Anammox cultivation in a closed sponge-bed trickling filter. In *XI Anaerobic Conference. XI DAAL Latinamerican Symposium on Anaerobic Digestion. November 24-27, 2014. Havana, Cuba.*

Journal articles

Sánchez Guillén J.A., Yimman Y., Lopez Vazquez C.M., Brdjanovic D., van Lier J.B., 2014. Effects of organic carbon source, chemical oxygen demand/N ratio and temperature on autotrophic nitrogen removal. *Water Science & Technology*, 69 (10), 2079-2084.

Sánchez Guillén J.A., Cuéllar Guardado P.R., Lopez Vazquez C.M., de Oliveira Cruz L.M., Brdjanovic D., van Lier J.B., 2015. Anammox cultivation in a closed sponge-bed trickling filter. *Bioresource Technology*, 186, 252-260.

Sánchez Guillén J.A., Jayawardana L.K.M.C.B., Lopez Vazquez C.M., de Oliveira Cruz L.M., Brdjanovic D., van Lier J.B., 2015. Autotrophic nitrogen removal over nitrite in a sponge-bed trickling filter. *Bioresource Technology*, 187, 314-325.

Sánchez Guillén J.A., Lopez Vazquez C.M., de Oliveira Cruz L.M., Brdjanovic D., van Lier J.B., 2016. Long-term performance of the Anammox process under low nitrogen sludge loading rate and moderate to low temperature. *Biochemistry Engineering Journal*, 110, 95-106.

Articles in preparation

Almedia P.G.S., Sánchez Guillén J.A., de Oliveira Cruz L.M., Lopez Vazquez C.M., Brdjanovic D., van Lier J.B., 2016. Anammox biomass adaptation under low temperatures: a model-based study.

Netherlands Research School for the
Socio-Economic and Natural Sciences of the Environment

DIPLOMA

For specialised PhD training

The Netherlands Research School for the
Socio-Economic and Natural Sciences of the Environment
(SENSE) declares that

Javier Adrián Sánchez Guillén

born on 22 January 1964 in La Chorrera, Panama

has successfully fulfilled all requirements of the
Educational Programme of SENSE.

Delft, 21 November 2016

the Chairman of the SENSE board

Prof. dr. Huub Rijnaarts

the SENSE Director of Education

Dr. Ad van Dommelen

The SENSE Research School has been accredited by the Royal Netherlands Academy of Arts and Sciences (KNAW)

KONINKLIJKE NEDERLANDSE
AKADEMIE VAN WETENSCHAPPEN

The SENSE Research School declares that **Mr Javier Sánchez Guillén** has successfully fulfilled all requirements of the Educational PhD Programme of SENSE with a work load of 38.4 EC, including the following activities:

SENSE PhD Courses

- Environmental research in context (2012)
- Research in context activity: 'Co-organising and coordinating the UNESCO-IHE PhD Symposium, Delft' (2012)
- Anaerobic wastewater treatment (2014)

Other PhD and Advanced MSc Courses

- Course on principles of anaerobic wastewater treatment, Wageningen University (2011)
- Modelling of wastewater treatment processes and plants, UNESCO-IHE Delft (2012)

Management and Didactic Skills Training

- Supervising four MSc theses (2012-2014)
- External reviewer for the journals 'Water Science and Technology' and 'Water Research' (2015-2016)

Oral Presentations

- *Cost-effective municipal wastewater treatment by coupling of UASB and Anammox reactors.* UNESCO-IHE PhD Symposium, 1-3 October 2012, Delft, The Netherlands
- *Effects of organic carbon source, COD/N ratio and temperature on Anammox organisms.* WEF/IWA Nutrient Removal and Recovery 2013, 28-31 July 2013, Vancouver, Canada
- *Low-cost Anammox cultivation in a closed sponge-bed trickling filter.* UNESCO-IHE PhD Symposium, 23-25 September 2013, Delft, The Netherlands

Poster Presentation

- *Effects of Organic Carbon Source, COD/N ratio and Temperature on Anammox Organisms.* World Congress on Anaerobic Digestion: Recovering (bio) Resources for the World, 25-28 June 2013, Santiago de Compostela, Spain

SENSE Coordinator PhD Education

Dr. ing. Monique Gulickx